# Putting Workfare in Place

# RGS-IBG Book Series

The *Royal Geographical Society (with the Institute of British Geographers) Book Series* provides a forum for scholarly monographs and edited collections of academic papers at the leading edge of research in human and physical geography. The volumes are intended to make significant contributions to the field in which they lie, and to be written in a manner accessible to the wider community of academic geographers. Some volumes will disseminate current geographical research reported at conferences or sessions convened by Research Groups of the Society. Some will be edited or authored by scholars from beyond the UK. All are designed to have an international readership and to both reflect and stimulate the best current research within geography.

The books will stand out in terms of:
- the quality of research
- their contribution to their research field
- their likelihood to stimulate other research
- being scholarly but accessible.

For series guides go to www.blackwellpublishing.com/pdf/rgsibg.pdf

# Putting Workfare in Place

## Local Labour Markets and the New Deal

Peter Sunley, Ron Martin
and Corinne Nativel

**Blackwell** Publishing

© 2006 by Peter Sunley, Ron Martin and Corinne Nativel

BLACKWELL PUBLISHING
350 Main Street, Malden, MA 02148-5020, USA
9600 Garsington Road, Oxford OX4 2DQ, UK
550 Swanston Street, Carlton, Victoria 3053, Australia

First published 2006 by Blackwell Publishing Ltd

1   2006

Library of Congress Cataloging-in-Publication Data

Sunley, Peter.
   Putting workfare in place : local labour markets and the new deal / Peter Sunley, Ron Martin and Corinne Nativel.
      p. cm. — (RGS-IBG book series)
Includes bibliographical references and index.
   ISBN-13: 978-1-4051-0785-3 (hard cover : alk. paper)
   ISBN-10: 1-4051-0785-5 (hard cover : alk. paper)
   ISBN-13: 978-1-4051-0784-6 (pbk. : alk. paper)
   ISBN-10: 1-4051-0784-7 (pbk. : alk. paper)
1. Youth–Employment—Government policy—Great Britain. 2. Welfare recipients–Employment—Government policy—Great Britain. 3. Labor market—Great Britain. I. Martin, Ron. II. Nativel, Corinne. III. Title. IV. Series.

   HD6276.G72S868 2005
   331.3′412042′094109049—dc22

                                                    2005013100

A catalogue record for this title is available from the British Library.

Set in 10/12 plantin
by SPI Publisher Services, Pondicherry, India

For further information on
Blackwell Publishing, visit our website:
www.blackwellpublishing.com

# Contents

# Series Editors' Preface

The RGS/IBG Book series publishes the highest quality of research and scholarship across the broad disciplinary spectrum of geography. Addressing the vibrant agenda of theoretical debates and issues that characterise the contemporary discipline, contributions will provide a synthesis of research, teaching, theory and practice that both reflects and stimulates cutting-edge research. The Series seeks to engage an international readership through the provision of scholarly, vivid and accessible texts.

Nick Henry and Jon Sadler
RGS-IBG Book Series Editors

# Preface

The move towards workfare policies represents a fundamental change in the welfare states and labour markets of many industrialised countries. Such a shift represents a process of activation in which the receipt of benefits and assistance are made conditional on the active fulfilment of job search and other work-focused obligations. Across the industrial world, politicians are identifying such policies as the solution to the entrenched problems of worklessness that have plagued their economies during the last few decades. Within Western Europe the UK has led the way in the adoption of workfare and the New Deal for Young People has been at the forefront. Buoyed by favourable national economic conditions since 1998, this flagship New Deal programme has been held up as a model to be emulated. While it has received much abstract and aggregate attention, there has been relatively little research into the uneven geography of the New Deal. The aim of this book is to look beneath the model and understand how this set of active policies has had quite different challenges and impacts in different local and regional labour markets. It attempts to contribute to the understanding of the role of geography in the constitution of labour markets, and to highlight the need to incorporate such understanding in order to construct effective and efficient policy interventions.

The geographical concentration of unemployment and worklessness has become one of the most problematic and stubborn features of the UK labour market. The book aims to consider how far the New Deal has been able to respond to and resolve this problem. How far have local flexibility and policy decentralisation allowed the programme to address dramatic differences in local labour market contexts? Despite the complexity of local outcomes, the book argues that the spatial variation in the New Deal tells a clear and systematic story in which the policy typically works more effectively in more dynamic and tighter local labour markets.

The geography of non-work is not a problem that has been virtually eliminated. Instead, the limitations and imbalance of supply-side active labour market policies, focused on raising individual employability, are most apparent in distressed local labour where there is less opportunity to find rewarding and stable job opportunities. The book discusses some of the implications of this finding for the idea of a new contract between unemployed individuals and the state. It outlines some of ways in which the local responsiveness of the policy could be improved, and some of the possible means of raising the demand for labour in depressed local areas. The need to do so remains pressing.

The research for this book was funded by the Economic and Social Research Council's Grant R000237866 (*The Geography of Workfare: Local Labour Markets and the New Deal*) and we would like to thank the ESRC for their financial support. We would also like to thank the many Jobcentre Plus (formerly Employment Service) officials, other local labour market agents, programme participants and employers who helped by providing information. We are especially grateful to those individuals who agreed to be interviewed in Cambridge, Edinburgh, Tyneside, Birmingham and North London. An earlier version of Chapter 3 was published in the *Transactions of the Institute of British Geographers*, 2001, NS Volume 26, pages 484–512, and an earlier version of Chapter 6 was published in *Environment and Planning: Government and Policy*, 2002, Volume 20, pages 911–932. We are grateful to the editors and referees of these journals. We would like to thank Tim Aspden and Bob Smith in the Southampton Cartography Office and Philip Stickler and Owen Tucker in Cambridge for their help in producing the figures. Finally we would like to acknowledge the late Pam Spoerry who drew many of the original maps for this project and provided much good-natured, professional help.

Peter Sunley
Ron Martin
Corinne Nativel
Southampton, Cambridge and Paris, August 2004.

# Tables

# Figures

# Chapter One

# Locating the New Deal

## Reforming Welfare and Redrawing Responsibility

There is little doubt that welfare states across the industrialised world are facing a set of unprecedented pressures and challenges. The globalisation of capital and trade, together with the intensification of global competition, have raised profound questions about states' fiscal capacities and the optimum levels of public spending and taxation. Technological change has been widely blamed for increasing levels of poverty and exclusion among unskilled and poorly educated groups, and the ageing of demographic profiles has raised serious questions about the viability of public pension schemes and welfare services. On top of all this, welfare states have suffered a relentless barrage of criticisms from neoliberal theorists accusing them of being thoroughly inefficient and counterproductive. European welfare states, in particular, have been targeted as sources of economic rigidity and have been charged with promoting social equality at the expense of employment growth. There is no doubt that welfare states are under stress.

But this does not mean that the welfare state is about to disappear. In fact, there is a widespread consensus that welfare states have shown remarkable resilience and continuity (Pierson, 2001; Taylor Gooby, 2001; Swank, 2002; Huber and Stephens, 2000). Neither does it mean that all welfare states are converging on a single model of residual or minimal welfare (Cochrane et al., 2001; Liebfried, 2001; Scharpf, 2001; Swank, 2002; Huber and Stephens, 2000). Conservative-corporatist, social democratic Scandinavian and liberal minimalist welfare state regimes continue to follow different trajectories, albeit with some complications and common trends (Esping Andersen, 1996; Taylor Gooby, 2001; Hall and Soskice, 2001).

But the absence of convergence certainly does not imply stasis. Political and institutional changes may mean that further reforms to European welfare states may be much more fundamental, and that the past will be of little help as a guide to the future. Taylor Gooby (2001) identifies two common new priorities. The first is *cost containment*, which means that no government is responding to pressing health care, pensions and labour market problems simply by increasing taxes, social contributions and spending. Increased capital mobility has reinforced the constraints of tax competition, and public sector deficits are avoided because of the discipline of the financial markets. In short there is a new emphasis on the need to contain and improve the efficiency of welfare state spending. The second is the acceptance of a *competitiveness imperative*. This asserts that welfare systems should primarily be oriented to sustaining economic performance and generating economic efficiency.[1] Thus social policies are no longer regarded as a distinct sphere, but are evaluated in terms of their interaction with economic policies and priorities (Giddens, 2000).[2] Cox (1998a, 1998b) and Lister (2002) perceive two similar common European trends in welfare provision. The first is *residualisation*, including a narrowing of entitlement and a greater targeting of social assistance, together with a move towards encouraging personal responsibility through the use of private welfare provision. The second is a mounting emphasis on making welfare rights conditional on the fulfilment of *citizenship obligations* and paramount among these obligations is the willingness to perform paid work.

> Social insurance and other benefit programmes are moving away from soli-
> daristic principles and becoming more achievement oriented. The notion of
> citizenship as the basis of an individual's claim to support is changing. There
> is an increasing demand that citizens recognise their obligations when they
> demand their rights (Cox, 1998a, in Lister, 2002).

As Lister (2002) explains, this new conditionality reflects the confluence of several streams of thought, including a reasserted Protestant work ethic and discourses of social exclusion that see paid work as the key route to social inclusion and full citizenship. In English-speaking states it also to some degree the consequence of powerful neoliberal and communitarian critiques of the consequences of welfare assistance.

During the last couple of decades, welfare debates have been suffused by a renewed moralism, which criticises bureaucratic welfare provision for inducing a weakening of personal responsibility and civic obligations. Deacon (2002) identifies several different perspectives which have contributed to this vision. The first is the *authoritarianism* of Murray (1984) and others which attacks the welfare state for creating perverse incentives which encourage declining levels of paid work, family breakdown and the creation of an underclass of welfare dependants. In this view, claimants are rational and self-interested individuals who respond rationally to perverse incentives. More generous benefit levels have created new levels of dependency and failed to validate individuals' responsibility for the consequences of their behaviour. The second perspective is the *new paternalism* of Mead (1997) that argues that welfare claimants are not best understood as rational, competent and functioning individuals whose behaviour is guided by incentives. Instead, they are dutiful but defeated actors whose culture condones self-destructive behaviour. Paid work should therefore be enforced and acceptable jobs should be broadly defined. A further perspective has been termed *responsive communitarianism* and argues that individuals in modern industrial societies have become too atomistic and have lost sight of the benefits of social responsiveness and commitment (Deacon, 2002). Rather than forcing and coercing compliance, governments should persuade people through moral arguments about the need to actively exercise personal responsibility and recognise their duties to their communities (Etzioni, 1999).

However, it is not just conservatives and communitarians who have emphasised the need to reform welfare and restore personal responsibility. Similarly, liberal egalitarian philosophers and authors arguing for a reconstruction of social democracy have accepted the need to take measures to encourage personal responsibility. For example, Dworkin (2000) argues that liberal egalitarian theories of justice have in the past tended to ignore personal responsibility. The first principle of Dworkin's theory of social justice is equal concern for all citizens, but the second is *special responsibility* – one person has final responsibility for the success of an individual human life – the person who's life it is. In principle, he argues, individuals should be relieved of consequential responsibility for those unfortunate features of their situation that are brute bad lack but not from those that should be seen as consequences of their choices. In this view, welfare programmes need to enforce rather than subvert proper principles of individual responsibility by being *endowment insensitive* but *ambition sensitive*, that is, in order to be just the resources provided should be sensitive to choices but insensitive to those circumstances attributable to (mis)fortune. A combination of collective and personal responsibility, in his view, represents the basis of a political 'Third Way'.

Giddens (1998, 2000) advances some very similar arguments. One of the defining characteristics of his 'Third Way' is a new social contract between the state and citizens based on the theorem of 'no rights without responsibilities'. In his view, the welfare state should be replaced by a 'social investment state' providing a degree of equality of opportunity primarily by investing in human capital and education.[3] He also argues that welfare restructuring should respond to moral hazard and perverse outcomes. 'It isn't so much that some forms of welfare provision create dependency cultures as that people take rational advantage of opportunities offered. Benefits meant to counter unemployment for instance, can actually produce unemployment if they are actively used as a shelter from the labour market' (1998, p. 115)

While these arguments clearly have very different implications and shortcomings, they share a critique of egalitarian welfare theory, together with the approaches typical of the 'old Left', for portraying benefit recipients simply as passive victims of forces beyond their control. This, it is argued, places too much emphasis on collective or state responsibility for welfare and not enough on personal agency and initiative, leading to perverse outcomes and moral hazard. This discursive theme has had a strong influence on the direction of Clinton's welfare reforms in the USA and has also guided New Labour's welfare reforms in the UK. The desire to redraw the boundaries between collective and personal responsibility has been one of the main motivations underlying the growing popularity of active labour market policy (ALMP) as a primary mechanism for reconfiguring the nature and operation of the welfare state.

## Activating Labour Market Policy

The 1990s witnessed a growing enthusiasm among economists and policy-makers for active labour market programmes. The term is applied to a wide-ranging set of measures designed to 'actively' intervene in the labour market in order to improve its functioning and efficiency, usually by introducing programmes for the unemployed (Calmfors, 1994). It is distinguished from 'passive' labour market policy, which essentially involves the payment of unemployment and other related benefits, and is often described as a 'safety net'. Active labour market policy includes *job matching* and placements services designed to improve the matching process between vacancies and job seekers; *labour market training* in order to improve the employability and skills of the labour force, and particularly those of the unemployed; and *job creation* schemes either by means of public sector employment or the payment of *recruitment subsidies* to private sector employers (Calmfors and Skedinger, 1995). In practice the distinction between active and passive measures is often blurred as no benefit systems have ever been entirely unconditional (Robinson, 2000; King, 1995). ALMP spans both supply-side measures such as training and compulsory welfare-to-work, as well as demand-side measures such as employment creation and the provision of guaranteed jobs.

The spread of active labour market policy has been predicated on the acceptance of an institutional (welfare-system induced) theory of unemployment. High unemployment in the late 1980s and early 1990s, particularly in

Europe, was interpreted as a *supply-side problem* in which welfare states imposed rigidities on labour markets and thereby prevented them from adjusting to a fall in the demand for unskilled labour, due to globalisation and technological change (OECD, 1994). In this account, the unemployed lack both the necessary skills and motivations to reconnect with the labour market and indefinite benefits allow them to drift into long-term unemployment. In this context, it was argued that Keynesian macro-demand policies are of little help as they fail to reconnect the unemployed to the labour market, and merely serve to inflate the wages of those in employment. This structural institutional/supply-side interpretation has coalesced into a policy paradigm that has guided and constrained policy makers' understanding of labour market problems (Larsen, 2002). As part of this paradigm, generous passive unemployment benefits are criticised as responsible for high and persistent rates of unemployment. In a highly influential text Layard et al. (1991) reviewed unemployment in 19 OECD countries and argued that those that responded well to economic shocks had welfare systems which discouraged long-term unemployment by offering benefits of 15 months or less, followed by active help to the unemployed. Indefinite benefits, they argued, are not in the interest of able-bodied individuals as they create moral hazards and reduce the intensity of welfare-to-work. Benefits should have a 'reasonable value' but should be accompanied by a stiff 'work test'. They concluded:

> What we have in mind is the Swedish mixture, or 'employment principle' as they call it . . . This assumes that it should be normal for those who want work to have it. In other words, the proper way to acquire an income is by work rather than by a state transfer. Thus, benefits should be paid only for a transitional period. But there should be active help (and ultimately a guarantee of temporary work) to those who have difficulty getting work (p. 473).

In this account, the Swedish system provided a policy model in the way that it balanced rights and responsibilities. A legally guaranteed right to a temporary job after one year's unemployment can only be delivered if there have been strenuous preceding efforts to get into work or training.

On the basis of such views, active labour market policies have been allocated a key role in the policy response to long-term unemployment (OECD, 1994). The OECD (1999), for example, agrees that passive welfare systems created perverse incentives and signals that discouraged people from taking work and failed to provide the intensive support required to help those detached from the labour market. It argues that 'this dependency leads to a heavy financial cost that constrains other public investments'(p. 10). In contrast, active schemes are often described as self-financing, as it is argued that they may pay for themselves through the reduction of benefit payments. They are believed to have a macro-economic benefit. By reconnecting the long-term jobless to the labour market, or converting them to short-term unemployed, it is argued that intensive job search, training schemes and wage subsidies can hold down wage inflation and so allow a higher rate of non-inflationary economic growth (Layard, 1997a). Active labour market policies have thus been described as 'Third Way' measures in that they supposedly reconcile employment and equality by simultaneously increasing both employment rates and social inclusion.

In fact, the statistical evidence on the effectiveness of the different types of ALMP is more mixed and ambiguous than some of these claims admit (Robinson, 2000), and the case that such measures produce a higher rate of employment growth has been found to be weak (Calmfors and Skedinger, 1995; cf. Kraft, 1998). The outcomes of such policies are notoriously difficult to identify with any real certainty (Calmfors, 1994). The intensification of active labour market policies in Denmark and the Netherlands during the 1990s has been widely praised for cutting unemployment and producing 'employment miracles', and these policies have been held up as policy models for emulation elsewhere (Cox, 1998b; Auer, 2000). However, even here, the precise effects of active labour market measures are debatable, complex and hard to isolate from business cycle effects, with different evaluation techniques showing different results (see Van Oorschot and Abrahamson, 2003). Optimistic statements about policy success have tended to be advanced without systematic evaluation and documentation (see Larsen, 2002; Martin, 2000). As Larsen (2002, p. 718) suggests, 'In this common European euphoria for activation policies some of the trade-offs and problems revealed by the evaluations are often ignored.'

Political rhetoric constructs 'active' and 'passive' measures as opposites. As Gilbert (2002, p. 189) writes, 'The word *active* speaks of life's energy, whereas *passive* suggests a state of mild depression.' In truth, active programmes are not alternatives to 'passive' measures, which compensate those who lose out, and there is little evidence that active measures remove the need for old forms of social protection (Hills et al., 2001). It is hard to avoid the conclusion that the policy enthusiasm for, and faith in, ALMP and welfare-to-work have run ahead of their unequivocal empirical achievements, largely because of the way in which they resonate with the supply-side interpretation of unemployment and the associated emphasis on the need to restore personal responsibility and work obligations. Nonetheless, evidence from Germany and Sweden suggests that while active measures may not be able to create employment, they may nevertheless play a useful role in maintaining the skills, motivation and morale of the long-term unemployed (Clasen et al., 1998). Comparative evaluations report that subsidised private sector employment can be highly effective in promoting the transition from welfare to work (e.g. Ochel, 2004), although large-scale programmes may suffer from significant displacement effects.

As references to the Swedish system indicate, active labour market policies were initially associated with European social democratic and some corporatist welfare regimes. Indeed the Swedes are credited with inventing the active approach to the labour market during the 1950s and 1960s. Forslund and Krueger (1994) noted that expenditure on labour market policy in the 1980s was about 3 per cent of GNP in Sweden, 2 per cent in West Germany and less than 0.5 per cent in the United States. The minimal spending on active measures in the US indicated a liberal welfare regime where Left-wing parties had little power (Janoski, 1994). Indeed in the post-war years active labour policy was an anathema to liberal regimes. Since the late 1980s, however, things have changed and in the context of political emphasis on the need to construct 'worker citizens', liberal regimes have embraced a particular form of ALMP, which has been christened *welfare-to-work* or *workfare*.

Workfare is an elastic and controversial term but was initially understood to mean making social assistance conditional on the performance of employment (Solow, 1998). The term can be used specifically to refer to those schemes where claimants are made to work for benefits at rates below the prevailing market wage (Gray, 2004). However, the concept is now usually understood in a broader sense to indicate the programmes where participants are required, as a condition of income support, to participate in a wide variety of activities that increase their employability and employment prospects (Peck, 2001, p. 84). Some argue that workfare is only one component of 'workfarism', which is a system of regulation designed to condition and coerce benefit claimants into taking low-wage, flexible and insecure jobs, thereby supplying a contingent labour supply that reduces wage pressure and encourages employment growth (Peck and Theodore, 2000a, 2000b). In this broad definition, tax credits for the low paid can be seen as part of workfarism as they increase the incentives for claimants to move from benefits into low-paid employment. Typically, workfare is designed to tackle a perceived problem of 'welfare dependency' characterised by poor motivations and weak work ethics among the unemployed. It is also clearly designed to deter individuals from seeking social assistance by making its receipt conditional on obligatory activities.

Workfare was closely associated with the steps taken by the Reagan administration in the US to make its welfare regime harsher and leaner, and during the 1990s under Clinton's Democratic Party the concept proved highly influential. While State governments in the US had been experimenting with workfare schemes since the 1970s, the Personal Responsibility and Work Opportunity Reconciliation Act of 1996 created strong incentives for states to move welfare recipients into jobs. The Act introduced mandatory employment targets for the welfare recipients and offered states new freedom in designing their welfare benefit schedules and implementing welfare-to-work programs (Blank and Card, 2001). Federal aid which provided cash assistance to parents with children (AFDC: Aid for Families with Dependent Children) was replaced by a federal block grant or funding stream that can be used in a variety of ways (TANF: Temporary Assistance for Needy Families). However, in order to receive this funding, states must meet or exceed a rising target for the proportion of welfare recipients who work at least 30 hours per week. A lifetime eligibility limit for the receipt of federally funded benefits of five years for adults also pressured states to reduce their welfare bills. A whole new set of TANF welfare-to-work programs has subsequently been introduced and has been associated with falling caseloads. Indeed, over the course of the 1990s the caseload across the entire country fell by 50 per cent (Blank, 2000). As Blank and Card write, this apparent success has been aided by a buoyant labour market: 'In this extraordinarily favourable macroeconomic environment, most states were able to focus on redesigning and implementing new programs, with little concern for job availability' (2001, p. 2; also Ellwood, 2000).

Because of its associations with punitive, residual American strategies, workfare in Europe has been a highly politically charged term and, while observers have tended to use the word in a pejorative and critical manner, policy-makers have tended to avoid the word and use 'welfare-to-work' or 'activation policies' instead. However, this may be changing. Lødemel and Trickey (2001) identify two main types of definitions of workfare: aims-based definitions, which concentrate on the objectives of the programmes, and form-based definitions, which emphasise the character of the policy. They prefer a form-based definition; workfare signals a compulsory programme in which non-compliance with work carries the risk of lost or reduced benefits. In their view, aims-based definitions are at risk of simplification and neglect the compatibility of workfare with different ideologies. In this way, Lødemel and Trickey attempt to shed some of workfare's political connotations, and represent it as a set of tools, put to different use in a variety of welfare regimes. While this seems to neglect some of the common ideological currents behind the shift to active labour market policy outlined above, there are undoubtedly different types of welfare-to-work programme.

Torfing (1999) identifies 'offensive' workfare strategies and 'defensive strategies'. Offensive workfare is statist and produces benefits for both capital and labour by providing education and high-quality training for the unemployed. Defensive strategies, in contrast, are neoliberal attempts to lower unemployment benefits and aggressively move the unemployed back into work. In his view, Danish workfare has followed an offensive strategy, so that 'The Danish case undermines the myth that workfare is essentially neoliberal, punitive and bad' (Torfing, 1999, p. 23; also Etherington, 1998). Peck and Theodore (2001), likewise, distinguish between 'work-first', labour force attachment schemes, which compel individuals to take formal employment as soon as possible, and 'human capital' schemes which focus more on the provision of training and skills development. Gray (2004) prefers to distinguish between neoliberal workfare measures that push up the labour supply and depress wage rates, and Keynesian measures that act to reduce the labour supply by providing training opportunities and jobs in the non-profit sector, and by raising the demand for labour by subsidising private sector jobs. Barbier (2001) also distinguishes two ideal types of activation. The *liberal* type enhances individuals' relationships to the labour market, incites individuals to seek work, and provides matching services and short-term vocational training. The *universalistic* type, in contrast, provides extended services to all citizens, and guarantees high standards of living for the assisted and benefit levels close to minimum wages. While no welfare states fit these types exactly, Barbier and Ludwig-Mayerhofer (2004) suggest that the UK is close to the liberal type, Norway and Denmark are universalistic, and Germany and France occupy intermediate positions.

In summary, despite the different classifications, the common theme is that it is possible to distinguish between *social democratic and neoliberal activation* policies. Both tend to make job search obligatory, but the former tend to concentrate more on training, temporary public employment and providing qualifications, while the latter also attempt to increase work incentives and price the unemployed back into work by lowering wages. However, as Larsen (2002) argues, both approaches are primarily supply-side and tend to accept the institutional supply-side theory of unemployment outlined above. But it is clear that workfare, and active labour market policy more generally, can have different characteristics and different outcomes in different welfare regimes. There have now been several attempts to document the spread of activation policies across different states and to examine their reliance on social democratic or human capital approaches as against neoliberal work-first schemes. It is clear that it is not always easy to characterise national schemes, because they involve complex mixtures of different components, they include different forms of conditionality and activity, and they interact with other differences such as the degree of reliance on social insurance. Moreover, in recent years, work-first activation measures have been introduced within the contexts of quite different national welfare regimes. For instance both Norway and Germany have introduced schemes that are in some ways similar to US work-first approaches (Lødemel and Trickey, 2001; Gray, 2004). Where, then, do the UK's New Deal programmes sit within this complexity?

## Introducing the New Deals

New Labour's 'Third way' approach to welfare reform insists that rights should be matched by responsibilities and presents active labour market policy as a solution to welfare dependency. The rationale was first set out in the Green Paper *A New Contract for Welfare* (Department for Social Security, 1998), which argued that there is a need to end a welfare system that 'chains people to passive dependency instead of helping them to realise their full potential' (p. 9). The welfare state should not be a residual safety net but should be modernised to provide for people in ways that fit the modern world. Its welfare-to-work programme would 'break the mould of the old, passive benefits system' (p. 24). The Prime Minister, Tony Blair, claimed, 'We have started to change the culture: a new approach to rights and responsibilities. Just as this Government is committed to creating more opportunities for people, individuals have the responsibility to take up those opportunities' (DfEE, DSS, HM Treasury, 2001, v). The government has argued that it has led the way in Europe in its adoption of active labour market policy and that it now has some of the most successful policies. Welfare reform has been designed to create an active or 'enabling' state, which is designed to overcome unemployment, rather than passively ameliorate it, and to 'cut the costs of economic and social failure' by preventing the unemployed from drifting into long-term unemployment and inactivity (DfEE, DSS, HM Treasury, 2001). As the rhetoric goes, the safety net should be transformed into a trampoline (Driver and Martell, 1998). In Blair's words, 'this is a government doing exactly what it said it would – modernising the welfare state on the principle of work for those who can, security for those who can't' (Blair, 1999a).

As this sound bite suggests, the government has endeavoured to reinforce the value and central role of paid employment. For example, the Chancellor of Exchequer, and the then Secretary of State for Work and Pensions, argued that the introduction of a single agency for unemployment benefits and job placement (ONE)

> will enshrine the obligation that everyone has an obligation to help themselves, through work wherever possible. In return, the Government has an equal responsibility to provide everyone with the help they need to get back to work, when they need it, as well as making sure there is greater security for those who cannot work (HM Treasury-DWP, 2001a, iv).

Paid employment is presented as the best route to economic independence and social inclusion and obtaining a job is claimed to be the quickest way to exit poverty. The government has promised 'a crackdown on the workshy' as part of its 'work-first' philosophy. For instance, according to Brown and Darling:

Work is the best way to lift families out of poverty, to raise incomes, and to open doors. That is why we have adopted a work first approach in our welfare reforms, as part of our drive to end poverty. For people who cannot work, our priority is to provide greater security than in the past (HM Treasury and DWP, 2001a, iii).

Similarly, in the words of Nick Brown, Minister for Work, 'we examine people's options for work before awarding benefit – a work first culture' (DWP, 2002).

In addition to the 'modernisation' of delivery and management, this 'work first' reform has involved two main elements: make work pay, and welfare to work. First, in line with its emphasis on the centrality of paid work, the government has introduced measures to increase the financial incentives of working. These include the Working Families Tax Credit and Child Tax Credit designed to increase the incentives of single mothers and families with children to enter work (see Blundell, 2001) as well as the National Minimum Wage. Such policies have been designed to counter the unemployment and poverty-traps created by the UK's means-tested benefit system and restructure financial incentives to make work more financially rewarding for the low paid.

Second, it has also introduced welfare-to-work policies designed to compel and encourage welfare recipients to gain paid work. In total, six New Deal programmes have been introduced (see Millar, 2000a). The flagship policy, the New Deal for Young People (NDYP), introduced in January 1998, contains the strongest element of enforcement of an obligation to work.[4] Since its inception various other 'new deals' have followed. The New Deal 25 Plus for the Long Term Unemployed targets those over 25 years and consists of advice and support from a personal adviser, followed by education/training or subsidised employment. Participation in the advisory programme is compulsory, while further participation is not. The New Deal for Lone Parents consists of an initial interview, job search and in-work support, and participation is voluntary. The target group is lone mothers on Income Support for six months with a child aged over five years. The New Deal for Partners of Unemployed People is a smaller, non-compulsory programme offering access to the NDYP and advice and guidance from a personal adviser. The New Deal Programme for Disabled People is voluntary and once again offers the support of a personal adviser, whereas the New Deal 50 Plus targets those in the 50 and over age group receiving incapacity benefits, Job Seeker's Allowance (JSA) or Income Support for at least six months. It offers access to a personal adviser and those finding work can receive a one-year employment credit. There is now also a New Deal for the Self Employed, to help people set up their own businesses. Yet further, a New Deal for Communities seeks to tackle multiple deprivation in the most deprived neighbourhoods of the country.

The focus of this book is on the NDYP (New Deal for Young People). It is the largest New Deal programme with a budget of £3,150 million, funded by an ex post windfall tax on the privatised utilities.[5] It aims to increase the employability of the young long-term unemployed and to move them into work. Those 18–24 year olds claiming the Job Seekers' Allowance for six months are entered on to the programme. There follows a Gateway of four months in which clients are interviewed by a personal adviser and given help with welfare-to-work. For the majority of clients the Gateway is their only experience of the programme, as summary statistics show that around two-thirds of clients leave the programme before the end of the Gateway. But at the end of this period, those clients who have not found work have to choose one of four options. These include full-time education and training, a work placement with an environmental organisation, a placement with a voluntary organisation, or a six-month subsidised job placement (subsidised at £60 per week and up to £750 for the cost of training). In addition, assistance is available to help young people move into self-employment. On the subsidised job option clients receive a wage that must be at least equal to the subsidy received by the employer from the DWP. Employers agree to provide training and a job at the end of the six-month period, providing the young person is deemed to be suitable. The full-time education and training option is aimed at those with no basic qualifications, and clients on this option receive an allowance for up to a year. The voluntary sector and environment work six-month placements are the most workfare-like options in the programme. While on these, participants receive their benefit plus a small supplement. The environmental option has proved to be the least popular. In September 2004, for example, Department for Work and Pensions (DWP) statistics showed that of the 70,850 people on the NDYP, 9.4 per cent were in education and training, 4.7 per cent working in the voluntary sector, 3.4 per cent were in subsidised jobs and only 2.9 per cent were on the environmental option. All the rhetoric heralding the programme made the point that there would be no 'fifth option' of continuing on benefit. For those who reach the end of their option without getting a job, there is a further period of follow-through advice and guidance, after which an individual still unable to get a job returns to benefit claiming and possibly back on to the programme

Government officials and others have argued that the design of the NDYP drew lessons from both social democratic Europe (particularly Sweden) as well as the liberal USA (Giddens, 2002; Annesley, 2003). Most academic observers, however, have stressed that the key characteristics of the programmes, particularly the NDYP, primarily reflected the operation of a process of transAtlantic policy transfer (King and Wyckham-Jones, 1999;

Dolowitz, 1998; Peck and Theodore, 2001; Deacon, 2002). American policy ideas have been much more influential than European social models, as institutional and labour market similarities between the UK and USA facilitate easier policy transfer and imitation (Daguerre, 2004; Daguerre and Taylor Gooby, 2004). In addition, lessons were also learnt from the Australian Labour governments' work-focused welfare reforms implemented between 1983 and 1996 (Johnson and Tonkiss, 2002). Dolowitz (1998) argues that the British Conservative governments 1979–87 had already established a system of workfare, primarily through the introduction of compulsory job search, via the Job Seeker's Allowance (JSA), and youth training schemes. While Peck (2001) agrees that elements of workfare were put in place under the Conservatives, he argues that it was not until the New Deal that a systematic workfarist system was introduced in the UK. The process of policy transfer spun out of the close links between New Labour and the Democratic Party, which allowed an intangible transfusion of policy languages, rhetoric and styles.

> In the case of policy welfare reform, what have been imported from the US are not so much individual 'policies that work' but more general *political strategies* of reform management – focusing on issues of dependency, the virtues of work, and so forth – coupled with a selective reading of key policy lessons. (Peck and Theodore, 2001, p. 430).

These reform strategies have clearly drawn on the conservative and communitarian critiques of dependency outlined above. According to several commentators, New Labour's welfare reforms have drawn heavily on communitarian ideas, and especially the notion that government should exhort and persuade citizens to practise their obligations and show a proper sense of moral and civic duty (Driver and Martell, 1998; Deacon, 2002). In this view, these reforms are a quintessential piece of Third Way thinking in that they attempt to reconcile what have previously been seen as opposing political principles, emphasising both individual responsibilities and collective obligations. While in principle it may be ultimately impossible to reconcile the irresolvable tensions and counterposed opinions of the political left and right, in practice New Labour's policies have attempted to find and manage pragmatic trade-offs and compromises between these positions (Driver and Martell, 1998). Thus its reforms of the welfare state attempt to draw on and accommodate different values. As Deacon (2002) explains:

> What is important here is not that New Labour has produced a new combination of incentives, authority and moral exhortation. It is that this new combination represents an attempt to respond to conservative ideas about dependency without abandoning altogether the goal of greater equality (Deacon, 2002, p. 105).

Much academic work, however, has been sceptical of any attempt to produce such a combination or balance. It argues instead that workfare and enforcing work obligations have taken priority over, and cannot be reconciled with, the aim of increasing social inclusion and equality. Instead, regulationist theories suggest that a regime of Fordist-welfarism is in the process of being replaced by one of post-Fordist workfarism, in which work obligations are forcing people into a liberal style of economy, which is acting to lock people into poverty rather than lift them out of it (Peck, 2001). In this view, while the UK's welfare-to-work programme is path-dependent and contains both continuities and discontinuities, it is nevertheless part of a generalised shift away from 'welfarist principles of needs-based entitlement and universality to the workfarist principles of compulsion, selectivity and active labour-market inclusion' (p. 429). Thus the New Deals represent the establishment of a workfarist system based on *enforcing work while residualising welfare*.

This approach draws on Jessop's account of the Schumpeterian Workfare Post-national Regime (SWPR), which, he argues, is replacing the Keynesian Welfare National State, partly as a response to economic globalisation, the growing openness of economies and a capitalist search for renewed profitability (Jessop, 1994, 1999). The ideal typical SWPR is said to have four key characteristics. First, it demotes the promotion of full employment, in favour of promoting structural or systemic competitiveness. Second, it is a workfare regime in so far as it subordinates social policy to the aims of labour market flexibility and competitiveness, aiming to put downward pressure on the social wage and cost of international production. Third, the SWPR signifies post-national welfare governance in the sense that other scales of analysis have increased significance. Not only is power transferred upwards to supranational agencies,

> But there is a simultaneous devolution of some economic and social policy-making to the regional, urban and local levels on the grounds that policies intended to influence the micro-economic supply-side and social regeneration are best designed close to their sites of innovation (p. 356).

Fourthly, the SWPR is destatised in that non-state mechanisms and private network partnerships are more important in the delivery of policies. Jessop further argues that this ideal-type has subtypes and variant forms and that it should not be reduced to its neoliberal subtype, although most of the writing on this approach seems to concentrate on this version. This account of the SWPR has since been criticised as too determinist and functionalist and for failing to identify the agents and actors responsible for political change, implying an unhelpful necessity in the course of welfare change (Gray, 2004). More recent regulationist work, keen to avoid the charge of crude functionalism, has stressed the importance of political choice, institutional pressures and the animation of neoliberal ideology (Peck, 2001). Emphasis has shifted away from economic necessities towards the rapid cross-national transfer of neoliberal policy ideas and strategies (Peck and Theodore, 2001). Nevertheless, it is maintained that workfarist regulation shows a certain emerging fit with deregulated neoliberal economic growth as it mobilises a ready supply of workers for low-skilled and insecure entry-level jobs (Peck and Theodore, 2000a, 2000b). Welfare reform and labour market change, it is argued, have evolved into a symbiotic relationship and after decades of experimentation and political struggle there is a degree of post-hoc functional correspondence between the contingent labour market and the direction of welfare reform (Peck, 2001).

This form of analysis has been highly critical of the aims and outcomes of neoliberal workfarism. Largely on the basis of North American experience, Peck (2001) argues that regressive workfarism recasts welfare in terms of short-term job market flexibility and enforces labour market participation in an era of low pay, work insecurity and low-grade service employment. A defining characteristic is that it is preoccupied with the initial transition into the labour market and therefore directs participants into low-grade, high turnover jobs. The effect is to exacerbate churning and turnover, and erode working conditions, in the lower reaches of the labour market. Workfarism is also argued to be of most help the most employable so that it acts to reproduce pre-existing labour market inequalities. In this view, it achieves little in terms of the alleviation of poverty, skill shortages or structural unemployment. Finally, Peck (2001) argues work-first systems are predicated on specifically local labour market conditions such that their results vary with the status of the local economy. In his words:

> Indeed real questions remain about whether work-first systems, much of which were developed in suburban or exurban locations, will even work in troubled, inner-city labour markets or in economically lagging regions. In some senses the purchase of the work-first policy package is weakest in precisely those areas where effective welfare-to-work programs are needed most: high unemployment areas (Peck, 2001, p. 356).

Indeed, recent work on the rapid decline of welfare caseloads in the US has shown that most welfare recipients continue to reside in inner city areas and there is evidence of an important spatial mismatch between these areas and suburban areas of job growth (see Allard, 2002; Allard et al., 2003).

In response to such contradictions, the critique of workfarism suggests that it depends on a continual search for new 'policies that work', cross-local/national policy transfers and fast policy churning. The geography of workfarism is argued to be integral to this incessant process of search, experimentation and diffusion of best practice. It involves a turn to local-ized delivery and experimentation and, as a consequence, *national welfarism* is being replaced by a *local workfarism* (Peck, 2001, p. 361). Spatial un-evenness and local distinctiveness together with the level of discretion of local administrators are all argued to increase as economic and social policy are recoupled at the local scale, contributing to a growing spatial uneven-ness in modes of local labour governance.

The obvious question from our perspective is whether and to what extent the New Deal corresponds with this vision of the mechanics and failings of local workfarism. Does it suffer from the same limitations as neoliberal regressive workfare and does it signal that the UK welfare state is set on a path to residual welfare? On the one hand there are clearly central aspects of the UK's welfare reforms that can be legitimately described as neoliberal. The country's welfare regime is undoubtedly set in a mutually reinforcing relationship with its liberal economy, and partly as a result, there are important continuities between Labour's New Deals and the welfare policies of the preceding Conservative governments. The pro-grammes have been implemented in the context of a pro-market economy with a deregulated labour market. New Labour has been keen to emphasise the benefits of a flexible labour market, with the addition of certain min-imum standards, as well as a stable macro-economy marked by fiscal prudence. Moreover, the Blair governments have also continued to imple-ment an 'organisational settlement', which aims at the modernisation of welfare and public service delivery through the introduction of 'managerial technologies' and practices borrowed from the private sector (Clarke and Newman, 1997, 2001).

Thus there are certain 'institutional complementarities' between this liberal economy and welfare policy (see Huber and Stephens, 2000). The relatively low level of benefits and the growing use of means testing re-inforce the operation of fluid and deregulated labour markets by ensuring a supply of labour for entry-level jobs (Esping Andersen, 1990). The bedrock of labour market policy continues to be the Job Seeker's Allowance, as the Treasury proudly proclaimed:

> Over the last decade the benefit regime facing the claimant unemployed has become progressively stricter. As well as receiving a relatively low level of benefits by international standards, the claimant unemployed must demon-strate that they are actively seeking work as a condition for receiving social security payments (HM Treasury 1997, p. 33).

While the government has introduced more incentives to reward work, its introduction of a minimum wage and tax credits has been designed so as not to significantly increase the relatively low wage costs of employers.

Indeed, in some way tax credits may enable and encourage low wages. While the New Deals contain a mixture of different types of active labour market policy, including wage subsidies that attempt to raise demand for the target group, they are primarily a supply-side intervention. Furthermore, the skills demanded by liberal market economies tend to be general skills, supplied via the educational system, which enable people to move from employer to employer, rather than firm specific and vocational skills provided by co-ordinated training systems (Hall and Soskice, 2000). The welfare-to-work regime in the UK can be described as liberal in character because of its combination of low benefit levels, a high degree of compulsion and an emphasis on means testing and individual responsibility to work (Wright et al., 2004).

On the other hand, however, New Labour's welfare reforms are in some ways more complex and contradictory than the workfarist picture of a degradation of welfare rights, and shift to a minimalist welfare regime. These reforms have been hallmarked by ambivalence (Lister, 2001). While going with the grain of a liberal market economy, New Labour has also shown a commitment to reducing social exclusion and has also implemented some more social-democratic strategies (Driver and Martell, 1998). It has made a commitment to halve child poverty by 2010/11 and eradicate it by 2020 and has raised the value of some benefits. The UK's welfare state has always been a mixture of liberal and social democratic collectivist principles and while recent reforms have undoubtedly moved it further in a liberal direction, it continues to be a hybrid (Clasen, 2003). As Clarke et al. write:

> It may be that the British welfare regime is in transition from one peculiar hybrid to another...The emerging regime appears to combine strong neo-liberal principles with more residual social democratic concerns (overlaid by occasional corporatist influences from the EU) (2001, p. 105).

For example, several observers suggest that the New Labour government has carried out a policy of 'redistribution by stealth' and has kept fairly quiet about its improvements in the level of some benefits, and transfers of income to families with the lowest incomes, for fear of alienating its middle-class support (Lister, 1999). The government's recent emphasis on explicit poverty reduction and social inclusion may have been too narrowly focused on employment as the sole route to inclusion, but it has generated a greater concern with dynamics of poverty and lifestyle change (Hills et al., 2001). Welfare reforms certainly include an increased targeting of benefits and income support on the poorest, but selectivity can be progressive; paying universal benefits to relatively wealthy middle class individuals is not an inherently preferable and sustainable alternative.

Furthermore, while the New Deal programmes are based on work-first, supply-side interventions, they also include some efforts to provide more human capital and job options and to provide personalised support for recipients (Millar, 2000a). As Gray (2004) argues, on the one hand, the New Deals represent a forced intensification of job search with workfare (benefit-plus placements) to follow for those do not succeed. On the other hand, the Employer Option is designed to provide jobs with a market wage, albeit the minimum wage in practice, and the New Deal programmes also temporarily remove a number of people from immediate job search by providing off-the-job training. As Gray (2004, p. 169) points out, only a minority of young people on the programme have been allocated to benefit-plus placements (about 15 per cent). And in her words, 'Training on the New Deal is longer than in any other programme seen in the UK since the late 1980s, and represents a substantial increase in government training investment' (p. 183). Lødemel and Trickey's (2001) comparative study of the development of workfare concludes that the New Deal has much in common with Dutch and Danish approaches to activation, and represents a relatively strong human resource development approach offering more opportunities to participants than is provided in workfare schemes elsewhere. Although they also note that the level of out-of-work benefits in the UK remains low and the move to compulsion is also relatively marked.

In summary, the New Deals have been part of one of the most fundamental reforms of a European welfare state to date, and this unusual degree of restructuring has moved the British welfare state towards a more liberal regime (Taylor Gooby, 2001). At the same time the New Deal programmes are a hybrid so that they combine some work-first components with other human capital opportunities. To a certain degree they reflect the influence of a liberal communitarian ideology of social inclusion and some degree of continuing concern with equality of opportunity. What is unclear, however, is whether and how far this hybrid form of workfare works in the same locally fragmented and erosive way as regressive workfarism, and has also suffered from the same types of problems, contradictions and counter-productive effects. In order to evaluate these questions in detail, we need to look both at the New Deal's geography of delivery and at the extent to which its outcomes have varied across different local labour markets. The interactions between economy and welfare reform do not stop at the scale of the national economy. Liberal production regimes have certain generic structural characteristics, but one of these is that they are also differentiated across space and show marked and contingent regional and local variations. How welfare regimes interact with such differences also depends on how they are also ordered and structured across space (Pinch, 1997). In short, both production and welfare regimes have important geographies.

## Inadmissible Evidence! Geography and the New Deal

It has been argued that the geography of any labour market programme is driven by two sets of processes (see Peck, 1994; Jones, 2000; Martin, 2000; Martin and Morrison, 2003). Firstly, it reflects the outcome of a process of *governance* in which central state and local agencies interact, and allocate power, responsibilities and functions. These institutional and political processes will determine how scales of intervention are constructed, how much local autonomy in decision-making is exercised under the programme and the degree of (de)centralisation typical of any programme. Secondly, however, there is another set of processes that centre on the *interactions between the intended and unintended effects of the programme and local labour market conditions*, which shape local outcomes. Such effects will be contingent on local conditions and so tend to produce a spatial selectivity in which policy has uneven spatial effects and reinforces uneven development. The two sets of processes are interdependent, of course, as the degree of perceived spatial selectivity will depend on how institutional hierarchies and spaces are constructed and operated, and, in turn, the geography of interaction may affect the understanding of policy outcomes and hence influence the power and influence of different policy groups.

Discussions of welfare-to-work and the New Deals have, in general, included many more references to the first set of processes. The main reason for this is that the introduction of active labour market policies has often been associated with a decentralisation of policy delivery. Responsibility for the delivery and often the design of active policies is often allocated to regional and local administrations within particular states. Reviewing employment and labour market policies, the European Commission (2000) noted that:

> Almost all Member States are decentralising the implementation of policies decided and financed at national level. Although this is done mainly through local public employment services (PES), there is an increasing tendency to build working partnerships with different types of local actors (p. 16).

According to the OECD (2001), active labour market policy works best and meets its aims more effectively when implemented with a local dimension (also Campbell, 2000; Finn, 2000). In Chapter 6, we explain this view and examine the purported advantages in detail. This emphasis on the benefits of decentralisation clearly had some influence on the design of the NDYP. The rhetoric introducing the programme suggested that it would involve a strong degree of decentralisation. Sir Peter Davis, Chair of the New Deal Taskforce, stated that the local partnerships would have as much discretion and autonomy as possible. The local partnerships delivering the NDYP would, it was suggested, be able to design policies tailored to their local economic circumstances (Education and Employment Committee, 1998).

The significance of this in theoretical terms is that it appears to conform with the argument that the spatiality of welfare governance is undergoing profound change. As we have noted, it is claimed that the development of Schumpeterian workfarism involves a process of a 'hollowing out' of the national welfare state (Jessop, 1999). We examine in Chapter 6 whether these contentions conform with the actual experience of the New Deal. We argue that decentralisation has actually been quite limited and restricted by the centralist character of the programme and its management structure. Moreover, we argue this is not simply an outcome of the path dependent nature of the highly centralised and monolithic UK welfare state. It also reflects the imposition of a new managerial regime that deliberately creates a restricted degree of local discretion within centralised controls, in order to ensure that local units compete to achieve central targets in the most efficient manner. Thus the complex changing institutional spaces surrounding the New Deal reflect not only the partial implementation of some lessons about decentralisation gleaned from the international transfer of best practice; they also reflect some of the 'managerial technologies' of the new public managerialism, through which the central state monitors, controls and audits the activities of local public institutions (see Clarke et al., 2001; MacKinnon, 2000).

In contrast to these debates, the second set of geographical processes – the local interactions between programme effects and conditions – has received much less attention. It has only made a minor appearance in the official evaluations of the New Deal, most of which focus on particular parts of the programme or on the aggregate and macro-economic outcomes. For example, it was not until 2002, after publishing several reports on the New Deal, that the Employment Select Committee of the House of Commons admitted that there may be significant geographical variations in outcomes. There has been relatively little analysis of local interactions and geographical contexts in commentaries on the New Deal. In some ways this was surprising. For example, there was already a considerable body of evidence from the US which indicated that workfare programmes were more effective in buoyant local labour markets (Newman and Lennon, 1995; Oliker, 1994; Friedlander and Burtless, 1995; Gueron and Pauly, 1991; Jensen and Chitose, 1997; Peck, 2001).

There have been both economic and political reasons for this lack of analysis of the geography of processes and results. First, as we have seen, the economic theory behind the New Deal explained long-term unemployment as the outcome of passive unemployment benefits and the loss of motivation and skills resulting from a detachment from the formal labour market. The issue was that individuals lacked employability, which was understood as referring to the skills and assets possessed by these individuals. This is a classic example of the way in which the discursive construction of a social problem determines the character of the policy response. Long-term unemployment was diagnosed primarily as an individual supply-side issue, so that local variations in employment growth and the demand for labour simply could not figure as important problems to be addressed (Turok and Webster, 1998; Peck, 1999). For fear of legitimising a lack of motivation and job search among the unemployed, HM Treasury has repeatedly and resolutely argued that there are no local job gaps or demand deficiencies causing shortages of vacancies. While there are pockets of multiple and complex problems, 'they do not face a simple lack of jobs' (2000, p. 7). So, according to the Treasury:

> The problem of Britain's most deprived areas is not necessarily a lack of jobs – in almost every case, these areas sit alongside, and within travelling distance of, labour markets with high levels of vacancies. People need to be equipped to take advantage of those opportunities. The Government therefore needs programmes to increase the employability of people in deprived areas, alongside those aimed at regenerating those communities, so that people from deprived areas can access and fill the vacancies that exist near to where they live (2000, p. 1).

While admitting that there remain long-standing differences in labour market opportunities between regions, in this 'Treasury view', the labour market is 'more balanced across the country, at a regional and sub-regional level, than at any time since the 1970s' (p. 9), and the decline of regional unemployment differentials 'seems to go well beyond' normal cyclical behaviour. The economics surrounding the New Deal has consistently focused on the health of the national economy, assuming that regional and local differences in unemployment are small and declining. In geographical terms, the tendency was to insist that a 'rising tide would lift all boats'. Such was the preoccupation with ensuring a stable macroeconomic context that local variations in economic growth faded into insignificance. Our review of local unemployment in Chapter 2 shows how partial and mistaken this was.

In addition, there have also been political reasons for this exclusion of the importance of geography to the NDYP. The character of the ideological support for the New Deal has oscillated between welfare dependency and a notion of social exclusion/inclusion. As we have seen, the former reinforces the focus on the behavioural and ethical deficiencies of individuals. At the same time, the latter assumes that there is a single national community in which the excluded could and should be integrated (Prideaux, 2001). Moreover, inclusion is to be based not on egalitarian outcomes but on the provision of the opportunity to work (Lister, 1999). The admission that the opportunity to work remains in fact geographically variable creates unwelcome complications, particularly if many of those areas with lower employment opportunities are in the traditional heartlands of Labour's electoral support. In general it seems that the Third Way's emphasis on mutual rights and obligations is uncomfortable with geography. If the right to get a job of a similar quality is significantly variable across space, and is partly dependent on where an individual resides, does this mean that responsibilities should also be variable according to location? It is not hard to see why a serious engagement with geographically uneven labour market conditions was ruled inadmissible evidence, as it confuses the notion of a national community based on shared moral obligations and weakens the pressure on the unemployed to find work.

However, notwithstanding the influence of 'Treasury view', we do not wish to overstate our case and suggest that the policy regime surrounding the New Deal has completely ignored the geography of labour market conditions and programme outcomes. Instead it has increasingly been argued that there are certain localities that need extra help. These areas are usually delineated as problem 'pockets'. In the Government's terms, more needs to be done to ensure that employment opportunities are available to those in 'remaining pockets of high unemployment and low employment':

> Unemployment has fallen in every region of the country. But there are still pockets of high unemployment, mainly in poor urban neighbourhoods, often within daily travelling distance of areas with high numbers of job vacancies, and people face difficulty finding work there (Department for Education and Employment, Department for Social Security and HM Treasury, 2001, p. 33).

Firmly 'on message' again, the Treasury and Department for Work and Pensions, describe how inactivity has become concentrated in certain disadvantaged areas:

> By the mid-1990s, inactivity had also become concentrated in certain areas of the country. These areas, where a large proportion of the population is outside the labour market, tend to be in large conurbations, or in areas that were in the past dominated by mining or heavy industry. Nonetheless, even in areas where there are still pockets of high worklessness it is often only a short travelling distance to areas where high numbers of vacancies and skill shortages can be found (2001a, p. 3).

Once again the construction of this local problem paves the way for the policy response: the exceptional and particular problems of localities are to be addressed by means of local zones and initiatives. These include Employment Zones, Action Teams for Jobs and, from April 2002, the StepUP scheme in 19 pilot areas of high unemployment. This is a clear example of New Labour's localism, which represents social problems as being confined to certain localised areas and amenable to area-based solutions (Mohan, 2000). Whether these additional locally focused initiatives represent an adequate response to the problem of locally uneven labour market conditions is discussed in Chapter 7, subsequent to our evaluation of the local impact of the basic New Deal across the country.

## Aims and Approach

The aim of this book, therefore, is to remedy this relative neglect of geography in the formulation and evaluation of the New Deal and to examine both the NDYP's institutional spaces and its local labour conditions and interactions in some detail. Our analysis is based in part on the findings of an ESRC-funded research project that used a combination of quantitative and qualitative methods. The research is founded on a nationwide mapping of some of the core performance measures and statistics made available by Jobcentre Plus (formerly Employment Service) in order to identify regional and local variations in New Deal outcomes, and these results are mainly presented in Chapters 3 and 4. In addition the book draws on a more intensive approach of five case study Units of Delivery (UoDs), selected to provide areas with contrasting labour market characteristics and also to ensure a north–south representation. A 'top-down' approach consisting of five key stages was adopted for each of these five UoDs (Birmingham, Cambridge, Camden/Islington, Edinburgh, North Tyneside).

First, semi-structured interviews, and occasionally small-group meetings, were carried out with the relevant regional managers in the Employment Service. The interviews covered issues such as the nature of the regional labour market (both in terms of supply and demand), the institutional set-up and spatial boundaries of New Deal delivery units, and managers' perceptions of programme implementation and the performance at a regional level.

Second, semi-structured interviews were carried out with district-level managers and addressed issues relating to the size, make-up and management of the New Deal 'strategic partnership', the meaning of local Delivery Plans, the state of local vacancies, the nature of the client group and internal management operations (such as marketing strategies, caseloads, and performance targets). In each of the five offices, two New Deal personal advisers were interviewed to explore their local labour market knowledge, personal case management practices, perceptions of clients' expectations and the obstacles faced when trying to direct their clients to employment and training opportunities.

Third, strategic partners from private, voluntary and public sectors were identified and contacted, and interviews arranged with an average of five partners in each UoD. Respondents included representatives from statutory agencies such as the former Training and Enterprise Councils, local and regional development agencies, regional colleges, local authorities, the TUC, and voluntary sector and grass-root organisations. The issues discussed with the partners included their perceptions of the local labour market, their role within the strategic partnership and their relationship to other partners and, especially, to the Employment Service.

Fourth, local employers listed as involved in the NDYP were contacted. Many of these employers reported that they had not taken on New Dealers despite having signed up, and in many instances, particularly in large organisations, it was difficult to track the person within the company who had originally signed the agreement. Several employers were reluctant to take part in the survey, so that the pool of participating employers was narrowed. On average, 18 employers representing a variety of sectors of activity, sizes and length of operation were interviewed in each of the five case study UoDs. This small-scale survey of employers was based on structured interviews providing quantitative data (relating mainly to New Dealers' earnings and hours of work) and qualitative information (New Dealers' employability and work attitudes, and relationship to colleagues). The employers interviewed included public, private and voluntary sector employers who were at various stages of involvement. The respondents were proprietors, general managers, company directors or human resources/personnel managers. Further, three non-participating employers were also interviewed in order to explore the reasons for non-involvement, particularly in buoyant local labour markets where numbers of young people on the subsidised employment option were comparatively low. Additionally, 15 large UK employers responded to a postal questionnaire, but the number of respondents was obviously too small to allow for statistical analysis. However, we utilised the responses to supplement the qualitative information obtained through the structured interviews. But the quantitative analysis presented in Chapter 5 is restricted to the 91 employers who participated in the main survey and who reported that they had recruited young people through New Deal.

Finally, informal interviews were held with 20 young people who had taken part in New Deal in the case study areas. The issues discussed with New Deal participants included their experiences in the programme and their perceptions of local labour market opportunities and expectations. In total 216 individual respondents were interviewed as part of the project.

This book presents the main results of this project as well as subsequent and supplementary research. It argues that the NDYP represents a nationally specific and hybrid form of workfare that has shown distinctive institutional and labour market geographies. It tries to evaluate whether and to what extent it has adequately responded to the unevenness and diversity in local labour market conditions and governance; whether and to what extent it can be said to have met its targets in different types of local economy. The following chapter sets the scene by examining the local geography of worklessness and unemployment, including those labour market problems that the programme was designed to tackle. It explains some of the interactions between demand and supply-side processes underlying this geography. Chapter 3 examines the geography of the programme's core outcomes across the country and highlights some of the most important dimensions of variations in its performance. Chapter 4 discusses the local workings of the programme in selected case study areas in more detail. It explains how the programme encountered different local problems, including different types of low employability. Chapter 5 then turns to the provision of training and the work experience obtained via subsidised employment. It develops a typology of different types of employers involved in the programme and discusses how this related to local labour market conditions. Chapter 6 considers the degree of local decentralisation and autonomy actually implemented under the NDYP and comments on current promises to extend the local flexibility of the programme. Chapter 7 concludes by discussing the normative challenge posed to the welfare-to-work paradigm by significant geographical variations in labour market opportunity and outlines some of the ways in which welfare-to-work could be improved in order to cope better with difficult local economic conditions.

# Chapter Two

# The Geographies of Worklessness

## Introduction

There is now widespread agreement that over the past two decades or so the world of work in advanced economies has been undergoing an accelerating process of transformation. Since the end of the 1970s, a number of intersecting economic, technological, social and political developments have been gathering momentum which have already had profound effects on the nature, organisation and allocation of work (see Figure 2.1). The shift from an industrial to a post-industrial, or knowledge-based and service dominated, economy and society, a wave of information- and knowledge-based technological change, increasing globalisation, the decline of old societal traditions and the rise of new cultures and identities, and fundamental shifts in political ideologies and policy regimes have all combined to sweep away many of the old verities and certainties concerning employment opportunities, job security, occupational divisions, wage structures and welfare entitlements. Labour markets are now much more uncertain, fluid and risky, and employment opportunities and wage rewards much more uneven, than they were only 20 years ago. It is no exaggeration to claim that we are witnessing the advent of new worlds and new rules of work (Herzenberg, Alic and Wial, 2000; Reich, 2001; Benner, 2002). For example, according to Reich (2001) the three key rules that underpinned the world of work that developed and flourished in industrial nations like the USA and UK from the late-nineteenth century through the twentieth, are now in retreat as we move to a 'new economy', to wit: the end of steady work, the necessity of continuous effort, and widening inequality.

Steady work – the security of a job (often for life) with regular and predictable pay – is disappearing for large numbers of workers, at all levels in the occupational and skill hierarchy. The new precariousness or 'risk' in

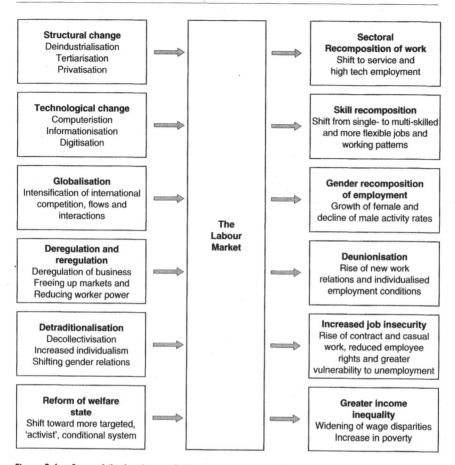

**Figure 2.1**   Some of the key forces of change and their labour market impacts.

the world of work (Beck, 1992; Elliot and Atkinson, 1998; Allen and Henry, 1997) is manifest in many ways. The rising tide of so-called 'flexible workers' – of part-time, temporary, freelance, casual and contract workers – has received most attention. But the job stability and pay of many full-time employees are also becoming increasingly volatile, as firms respond to the vagaries of the new competitive economic environment by readily firing and hiring staff, and by increasingly tying pay to individual performance (in the form of productivity, sales commissions, bonuses, shift payments, and the like). Added to this, the fact that most new jobs are now created by small firms (typically fewer than 25 employees), and that many of these are prone to short life spans, means that average job tenure in such businesses tends to be uncertain. Indeed, there has also been a growing debate about the future of employment and work: will there be sufficient jobs for everyone who wants to work? Much of the discussion of this question has revolved round two contrasting scenarios.

The first, and most pessimistic, is that which espouses a spectre of 'jobless growth'. The key argument here is that contemporary developments in technology, automated production and services, and global competition, are increasing the efficiency and productivity of labour to the extent that output can be expanded without additional labour inputs, indeed using significantly less labour. Some exponents of this view point to economic history for empirical support of their argument. Thus, for example, Dunkerley (1996) argues that we observe a common repeated pattern in employment growth as economies pass through successive phases of development and major sectoral change, with agriculture, then manufacturing, and possibly soon services, undergoing first employment expansion, followed, second, by employment contraction. Such writers point especially to the onset of deindustrialisation amongst several of the OECD economies from the late 1960s, and especially the 1970s, onwards, and to the millions of workers expelled from manufacturing in the last quarter of the twentieth century, partly due to technological change, partly to the rise of new international competition and the shift of manufacturing from the advanced economies to low-wage countries.

Further, according to these proponents of the 'jobless growth' thesis, there are now signs that the same process has since begun to spread to many of the service industries, as they too adopt new information-based technologies, and even into some of the 'new economy' high-tech sectors themselves (Lisbon Group, 1995). Thus the global computer and semiconductor industries, until recently major employers paying high wages, are now among the chief beneficiaries of their own efficiencies: as the 1990s wore on, so companies making silicon chips and computers turned from being beacons of employment creation to engines of job loss. Many financial and related business services experienced a similar job downturn. The call centres that expanded so rapidly in the late 1990s are now beginning to transfer operations to low-wage countries (especially India), like some sections of routine manufacturing before them. Everywhere job insecurity has crept up on the highly and technically skilled. Skills that are less than two decades old are being superseded by new skills whose life span will be considerably less than those they replace. Many of the people being made redundant by technology, structural change and automation, are only able to find new jobs at lower wages or in lower skilled work. Being a graduate no longer guarantees ready movement into a well-paid, secure job. While believers in the 'jobless growth' future acknowledge that the mass unemployment that characterised many OECD countries in the 1980s and 1990s has receded somewhat, they point to high and increasing levels of non-employment and economic inactivity as concealing large numbers of unemployed that do not figure in official counts of the jobless. The fear is that it now requires ever-increasing rates of unsustainable economic growth to maintain, let alone increase, employment.

The second perspective, the 'jobs miracle' view, advances a diametrically opposed prediction. Here it is argued that while new technologies may destroy old inefficient jobs, it simultaneously creates many more new ones in new firms and new industries. The stagnation and decline of employment in manufacturing is not seen as being problematic – or at least only in the short run – and even as a positive thing, since services and high-tech activity have generated large numbers of new jobs, more than enough to compensate for those lost in manufacturing. The problem, it is argued, is not a lack of jobs, but making sure that people have the skills to fill them. Indeed, according to some 'new economy' versions of this argument, a major 'paradigm shift' has taken place. As the result of rapid ongoing technological innovation, substantial productivity growth, the informationisation of the economy, the proliferation of personalised services, the flexibilisation of labour markets, economic liberalisation, and tax reductions, a new model of capitalism is seen as emerging in which both inflation and unemployment have been beaten and recession all but expunged. Adherents of this largely US-centred scenario point to millions of new jobs created in that economy over the past two decades or so, and to the long 1990s boom – one of the longest on historical record. While it is the case – certainly compared to the European Union, for example – that the USA has experienced a 'jobs miracle', many of those jobs have been poor ones, requiring few skills and paying low wages, with the result that income inequalities widened during the 1980s and 1990s (Krugman, 1994; 2002; Dickens and Ellwood, 2000). And the sudden downturn in the high-tech sector of the US economy in late 2000 (the bursting of the 'dotcom' bubble) suggests that in contrast to what has been claimed (Coyle, 2001), the so-called new economy is far from immune from recession and major job losses. The 'jobs miracle' may in fact turn out to have been something of a mirage.

Each scenario on its own is obviously too simplistic and too generalised a depiction of labour market trends. The reality is that different groups in the labour market face different prospects. For many – indeed, the majority – there are sufficient jobs available, albeit if many of those jobs are low wage, and have few if any career possibilities. But for a sizeable – and in some countries, growing – minority, the experience has been one of exclusion from the world of work. Indeed, one might argue that in many OECD countries a fourth rule of the post-war labour market – one not really highlighted by Reich – is also in retreat, or at least has become elusive, namely full employment. If there is a new world of work, there is also a new world of non-work. Furthermore, as we show in this chapter for the case of the UK, this issue of worklessness – of unemployment and other forms of non-work – has an explicit and non-trivial geographical dimension.

## The Unemployment Problem

The problem of non-work first emerged in the form of rising unemployment rates in the late 1970s and especially the 1980s. After the historically low rates of the post-war boom years, unemployment in almost all of the OECD countries began to rise sharply with the abrupt economic slow down following the impact of the 1973–4 OPEC oil-price hike. In 1973, some 10.5 million were officially classified as unemployed within the OECD area. By 1979, the time of the second oil price shock, this had increased to just over 18 million. The deep economic recession of the early 1980s then pushed the total up to 30 million. Thus in the space of just a decade, the unemployed had tripled (OECD, 1994). Strong economic growth during the 1983–90 period did bring some reprieve, but it failed to drive unemployment totals down to their 1970s levels, and with the return of recession in 1990–3, unemployment rose again to reach a peak of 35 million in 1995. Few countries escaped these developments, although the member states of the European Union were much more severely affected than the USA or Japan. Although unemployment rates across the OECD have ameliorated since the early 1990s, in some countries they have proved more resistant to improvement

There have been numerous explanations advanced for the rise and persistence of high unemployment across most of the advanced economies over the past two decades or so. Crudely put, these divide into demand-side and supply-side arguments. On the demand-side, one contention has been that the general slowdown of the global economy during the 1970s combined with the sharp recessions of the early 1980s and early 1990s, caused major shakeouts of labour and consequential persistently high rates of unemployment. Further, intersecting with slow overall economic growth, major structural changes in the demand for labour also took place. The shift of the economy from industry to services and technological activities has been accompanied by a progressive decline in demand for certain sorts of skills and occupations. Some variants of this explanation also point to the shifting international division of labour associated with this post-industrialisation of the advanced economies, as low-wage competition from developing countries has further undermined industrial jobs in the former. There has been considerable debate over the technology versus trade explanations (Wood, 1994; Lawrence, 1996).

Notwithstanding this dispute, the basic contention is that labour demand in the OECD countries has shifted from low-skilled to high-skilled jobs. At the same time there has been a long-term rise in the social demand for education and this has led to a growing supply of workers with higher-level skills – as measured by the educational attainment levels of the workforce. The problem has been that these two processes have not matched, so that the demand for low-skilled workers (relative to high-skilled jobs) has fallen more rapidly than the supply of such workers, so that large sections of the low-skilled workforce – especially prime age manual male workers, and unskilled young labour force entrants – have faced a major jobs gap, at the same time that many highly skilled and technological occupations have experienced a labour shortage. The evidence lends some support to this thesis, in that in several OECD countries unemployment rates for unskilled and manual workers have been higher than for more skilled workers (see e.g. the analyses by the OECD, 1994).

This tendency has been compounded by a distinct gender dimension to the shifting demand for labour. The decline of male manual, low-skilled and old-skilled jobs in industry has contrasted with the growth of female employment, particularly in service activities. In the UK, for example, the employment rate for working-age males declined from 95 per cent in 1960 to 75 per cent in 1993. It has recovered slightly since, but is still below 80 per cent (Figure 2.2). Over the same period, women's participation in the labour market has risen steadily: the employment rate for women was less than 50 per cent in the early 1960s, but had reached 70 per cent by the early 2000s. The increasing activity rates of women in the labour force reflects not only the growth of a range of service jobs that have attracted women into work (and which employers have targeted to women), but also a variety of socio-cultural factors, including rising educational levels. If for women the distinctive feature of the labour market of the 1980s and 1990s was expanding employment and participation, for men – both young men entering the labour market for the first time, and older, prime working age males – the distinctive feature was the high rate of long-term unemployment.

It is a well-established fact that the rate of long-term unemployment rises as overall unemployment increases. Unemployment is not simply a stock – the number of unemployed at any given time – but the complex product of flows into and out of unemployment and the duration of being unemployed. The more serious the deterioration in the labour market, the deeper is a recessionary downturn, or the more severe is structural decline in particular sectors of economic activity, the greater will be the inflows into unemployment and the more difficult it will be for those made unemployed to find a new job, so that the average duration of a spell of unemployment will also increase. Thus the rate of long-term unemployment – out of work for a year or more – will rise. In the 1980s it became fashionable, especially among monetarist economists, to argue that the rise in long-term unemployment over the second half of the 1970s through the 1980s itself

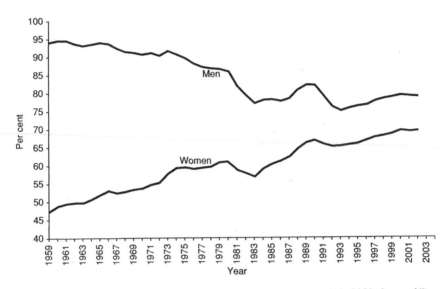

**Figure 2.2** Employment rates of working-age men and women in the UK, 1959–2003. Source: Office for National Statistics.

contributed to the problem of high and persistent unemployment by generating so-called 'hysteresis' effects, with highly adverse consequences for macro-economic stability and management.

The argument was that a host of supply-side factors and rigidities had combined to make labour less willing to search for work and employers less willing to hire, thereby pushing up the underlying 'equilibrium' or so-called 'natural rate of unemployment' (NUR) in the economy:

> so we may conclude that, aside from some notable examples, the secular shifts in unemployment which we have seen are driven by shifts in the equilibrium rate...so the next step is to discuss the factors which impact on the equilibrium unemployment rate...the variables which we might expect to influence equilibrium unemployment include the unemployment benefit system, the real interest rate, employment protection laws, barriers to labour mobility, active labour market policies, union structures and the extent of co-ordination in wage bargaining, labour taxes, and unexpected shifts in the terms of trade and trend productivity growth (Nickell, 2003, pp. 7–8).

According to Monetarist and New Classical theory, the 'natural rate of unemployment' has a major impact on the inflationary behaviour of the economy, for the NUR is that rate of unemployment at which the rate of inflation is constant (hence it is sometimes referred to as the non-accelerating inflation rate of unemployment, or NAIRU). If unemployment falls below the natural rate – for example as a result of a boost to aggregate demand in the economy – then this will tend to increase the rate of inflation. The higher the natural rate of unemployment, therefore, the more inflation prone is the economy, and the less effective are attempts to manage economic growth, for example through fiscal expansion. This interpretation of the rise of unemployment in the 1980s and its persistence through to the 1990s due to hysteresis effects associated with long-term unemployment became part of the conventional wisdom among many leading labour economists and macro-economists. Thus towards the end of the 1980s, Layard and Nickell (1987, pp. 148–149) referred to 'the pile-up of long-term unemployment' and asserted that 'If there were now a major economic recovery . . . it is most unlikely that long-term unemployment would fall at all rapidly, unless specific measures were taken to encourage employers to hire the long-term unemployed.' Similarly the OECD (1988, p. 28) argued, 'Long-term unemployment is now more serious than it was in the pre-recessionary 1979 period and is most unlikely to revert to the levels which prevailed at that time. This suggests that what we are observing is a ratchet-type effect.' Without doubt, this view played a critical role in the shift in policy circles away from the demand-side approach and passive benefits systems of 1950s, 1960s and 1970s towards the neoliberal supply-side 'make-work-pay' and 'workfare' policies of the 1980s and especially the 1990s, described in the previous chapter. Thus Layard's (1997a, 1997b) influential advocacy for workfare policies in the UK was based on the claim that reducing long-term unemployment (and by implication the 'natural rate of unemployment') would bring the macroeconomic benefit of lower inflation.

There are several versions, or theories, of the hysteresis or 'ratcheting-up' argument. According to the so-called 'duration- or state-dependent' thesis, the probability of exit from unemployment is dependent upon the length of time for which a person has been in the state of being unemployed. Spells of long-term unemployment render the workers affected increasingly difficult to re-employ and thus prone to further spells of (long-term) unemployment. This view can be traced back to the inter-war years (if not earlier), when, in the context of widespread mass unemployment, attention focused on the damaging effects on workers of long spells of joblessness. Thus Pigou (1933, p. 16) asserted that 'If a man is subjected to unemployment for a long period of time, injurious reactions on his industrial and human quality are almost certain to result . . . when opportunity comes again, the man, once merely unemployed, is found to have become unemployable.' On the one side, the longer a person is out of work, the more likely his or her skills begin to wane, or fall behind those required by employers. Further, so the argument goes, the long-term unemployed are likely to fall prey to welfare-dependency, especially if unemployment and related benefits are 'generous'. More seriously, the longer a person is unemployed the greater the danger of losing the motivation to work. On the other side, employers – especially under slack labour market conditions – are likely to discriminate against the long-term unemployed when hiring labour, in favour of workers with no history of repeated or long-term unemployment. In effect, the long-term unemployed can become stigmatised and discriminated against by employers. The 'state-dependent' or 'duration-dependent' idea was advocated in particular by Budd et al. (1988), Layard et al. (1991) and the OECD (1983, 1988). It has a very obvious policy implication, namely that it is overwhelmingly important to catch people early in each unemployment spell and give them work or training to prevent their loss of employability. This has been one of the considerations underpinning the development of the New Deals, and of 'active labour market policy' (ALMP) in general.

More contentious, however, is the 'benefit-dependency' argument. This theory is based on the idea that unemployment durations are higher when the ratio of benefits received by people out of work to net wages obtainable in work (the so-called 'replacement ratio') is higher, and/or the duration of benefits is longer. The claim is that the replacement ratio had drifted upwards over the 1970s and 1980s, as a result of increases in benefits, thereby making re-entry of the unemployed back into work – especially low-paid work – increasingly less attractive. This, it was argued, encouraged the long-term unemployed to become dependent on benefits, and to prefer worklessness to (low-paid) work. Accordingly, many commentators, especially those of a right-wing disposition, argued that unemployment benefits had became 'too generous' by the 1980s. Moreover, under the typical post-war model such benefits were automatic and 'as-of-right', with no requirement by recipients to undertake job training, work experience, or similar labour market participation, and this feature was also claimed to have encouraged 'benefit dependency'.

There is a basic implausibility about this idea in relation to the UK. To explain the huge variations in long-term unemployment since the 1970s there would have had to be commensurately large variations in replacement ratios and/or benefit durations. Up to the mid-1980s, these had not occurred, and the significant changes since then have been in the wrong direction to account for rising long-term unemployment. Even Layard et al. (1991, p. 258) themselves conceded that 'In Britain neither the replacement ratio nor the duration of benefits has altered much since the mid-1960s.' Thus it is difficult to attribute the rise and persistence of unemployment from the mid-1970s through to the mid-1990s to increasingly generous benefits. There is, in fact a good deal of direct British evidence that fails to find any convincing link between the two (e.g. Budd et al., 1988; Bean, 1994). Nevertheless, this type of account has had a major impact on British labour market policy.

Starting in the early 1980s there have been successive reductions in unemployment benefit levels and their duration. In part this has probably simply reflected the fact that it is an apparently easy way to save money, and an intuitive feeling that even if there is no evidence that attractive benefits might lead people to stay unemployed, there is a possibility that they might. However, it has also reflected a widespread belief that the fact that some countries – most notably the United States and Canada, and to a lesser extent Australia and New Zealand – had lower ratios of long-term to total unemployment which were attributable to their less generous benefit regimes. Making cross-country comparisons, and drawing policy implications from them, is difficult however, because of differences in the level and duration of benefits as well as in the eligibility criteria for their receipt, that is differences in the way that unemployment and especially long-term unemployment are defined and the ways in which benefits are administered (Webster, 1996, 1997). What differences do exist could at least be partly due to the effect of the benefit system on whether unemployment is officially classified as such, rather than to its effect on the real level of unemployment. This has recently become a major issue in the UK, as we shall discuss below.

Not only is the concept of a 'natural' or 'equilibrium' unemployment rate itself highly problematic and contentious (the early critique by Robinson, 1986, is still one of the most cogent), the various accounts that have sought to argue that the 'equilibrium unemployment rate' rose in the 1980s and proved resistant to reduction because of hysteresis effects associated with long-term unemployment have proved less than convincing. As the work of Webster (1997, 2003) has demonstrated, there has never been any problem of irreversibility in long-term unemployment and arguments and policies that identify hysteresis effects as the culprit are misguided. A rise in unemployment duration is a natural and unavoidable consequence of a rise in the overall unemployment level. The proportion of unemployment that is long term (more than a year) reverses itself with falling total unemployment (Figure 2.3). While this may take a year or so (in part because of the very definition of long-term unemployment), it does not permanently raise the underlying equilibrium or natural rate, as many theorists and policy-makers have assumed. In both the 1980s and the 1990s, the dramatic rise in

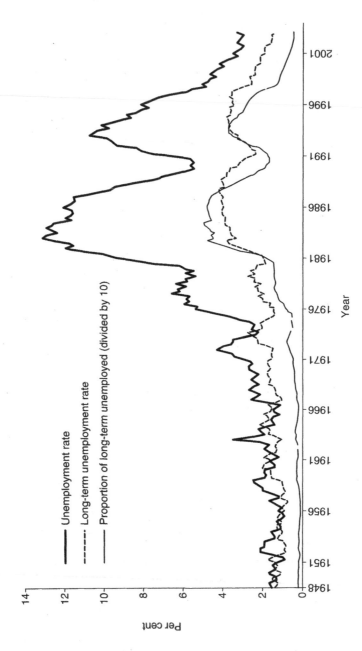

**Figure 2.3** Total and long-term claimant count unemployment, 1948–2003 (quarterly data, not seasonally adjusted). Source: Based on Webster (2003).

unemployment following the deep recessions of 1979–82 and 1990–2 was characterised by a corresponding marked increase in the long-term unemployment rate and the long-term unemployed as a proportion of all unemployed. Likewise, as the unemployment rate fell in the subsequent boom periods (1983–90 and 1993–2003), so long-term unemployment also fell. To be sure, high unemployment persisted much longer following the 1979–82 and 1990–2 recessions than it did in previous post-war economic cycles. But this was almost certainly due to the very intensity of the 1979–82 and 1990–2 downturns, the most severe since the inter-war years. The recent downturns were not just normal cyclical phenomena, but structural disruptions, involving major sectoral shifts in the economy, and resulting in substantial realignments and mismatches in the patterns of demand for and supply of labour.

If there is validity to the hysteresis argument it is this: the deeper and more structural a recession, the larger the numbers of workers thrown out of work, the larger will be the sections of employment that will have permanently disappeared, and the longer it will take for many of the jobless to find new employment, even when the economy picks up, since the new pattern of demand for labour will have changed in dramatic ways from the old. Simply put: the deeper the recession, the longer it takes for the labour market to recover and adjust, and the more protracted will be the average duration of unemployment. This need have nothing to do with the generosity or otherwise of unemployment benefits, though the negative 'duration-dependent' effects involving the erosion of skills, motivation and 'employability' among the long-term unemployed will be real enough.

There are, however, two issues that complicate this picture. As Figure 2.3 shows, according to the official claimant rate, unemployment in the UK appeared to fall sharply after the early 1990s: by 1997 – when the Labour government came into power – it had fallen back to around 5 per cent, or about half its peak in 1993. It has continued to decline, to levels and rates not seen since the early to mid-1970s. Yet this improvement has not gone uncontested. It has been argued that the official unemployment statistics do not tell the whole story about contemporary worklessness. It is now widely accepted that there are problems with official statistics on unemployment that make it difficult to separate involuntary demand-deficient from supply-side voluntary unemployment, let alone to measure or calculate the 'equilibrium' rate of unemployment. The official count of the number of unemployed is the result of an administrative decision that may not be totally immune from political considerations, and that can provide opportunities for the manipulation of 'headline' statistics. In the UK, the traditional measure has been the so-called 'claimant count', those individuals out of work and who are eligible to sign on for unemployment benefits. Since the early 1980s there have been numerous (some observers cite more than 30) changes to the eligibility criteria relating to the official 'claimant count', the majority of these restricting the numbers able to claim and hence officially counted as unemployed. The result has been mounting criticism of this traditional measure as an indicator of the 'real level' of unemployment in the UK (see Beatty et al., 1997, 2002), and growing pressure to adopt a more inclusive measure.

Allied to this, despite the fall in official unemployment the numbers of economically inactive – that is neither employed nor officially unemployed – has risen and remains at historically high levels. The decline in economic activity has been almost entirely a feature of the male population, and a distinctive feature of all major OECD countries. In 1970, for example, the male activity rate (employed or officially unemployed) in the EU was 92 per cent; by 1990 it had fallen to 79 per cent, and has since continued to decline. The decline in activity rates for men in OECD countries has been concentrated among the low-skilled, and has been most rapid for men with few educational qualifications, including new, young labour force entrants.

Clearly, not all inactivity reflects an underutilisation of human resources. Many women are not in paid employment nor unemployed, but are producing highly valuable services in the form of household work and the caring of children. Many young people may not be engaged in work but are instead investing in their own human capital in the form of education and training. In addition, some inactivity reflects a voluntary choice of leisure rather than work. But the growth of economic inactivity among men has a number of worrying features. It began to escalate precisely at the same time that official unemployment rates rose, from the late 1970s onwards. Since many of those who are inactive would no doubt like to work if suitable jobs were available, that is are 'hidden' unemployed, the growth in inactivity has almost certainly meant that official unemployment rates, however measured, have seriously underestimated the real scale of unemployment.

The measurement of 'real' unemployment is therefore of consequence. The definition usually used in international comparisons is the ILO (International Labour Office) measure, that is those, who during any given reference period, (1) had no employment, (2) were available to start work within the next two weeks, and (3) had actively sought employment at some time within the previous four weeks. A yet 'broader' definition is used in the European Labour Force Survey, namely the ILO definition plus those inactive members of the working age population who would like to work or are seeking work or who are available for work. In recent years the UK government has begun to move towards ILO definitions and Labour Force

Survey estimates, although the latter pose problems of small sample sizes when used to calculate local unemployment rates (so that averages across surveys are often used).

Using ILO definitions, there is clearly considerable variation in aggregate unemployment rates across the OECD countries (Table 2.1). Further, even in those countries where official unemployment levels have abated from their historical peaks in the 1980s and 1990s, inactivity rates have remained high: in many countries up to a third of the working-age population are economically inactive. In the UK, a quarter of the working-age population is economically inactive: this represents around 7 million individuals. The British government has tended to argue that unemployment and inactivity are not that closely related, and thus that inactive people should not be included in measures of unemployment:

**Table 2.1**    Unemployment, employment and inactivity across the OECD economies

| Country | Unemployment Rate 2001 | Unemployment Rate 2002 | Inactivity Rate 2001 | Employment Rate 2001 |
|---|---|---|---|---|
| Netherlands | 2.4 | 2.8 | 24.3 | 74.1 |
| Switzerland | 2.6 | 2.6 | 18.8 | 79.1 |
| Austria | 3.6 | 4.1 | 29.3 | 67.8 |
| Norway | 3.6 | 3.9 | 19.7 | 77.5 |
| Ireland | 3.8 | 4.4 | 32.5 | 65.0 |
| Portugal | 4.1 | 4.4 | 28.2 | 68.7 |
| Denmark | 4.3 | 4.2 | 21.8 | 75.9 |
| US | 4.8 | 5.6 | 23.2 | 73.1 |
| UK | 5.0 | 5.2 | 25.1 | 71.3 |
| Japan | 5.0 | 5.4 | 27.4 | 68.8 |
| Sweden | 5.1 | 5.0 | 20.7 | 75.3 |
| New Zealand | 5.3 | 5.3 | 24.1 | 71.8 |
| Belgium | 6.6 | 6.9 | 36.4 | 59.7 |
| Australia | 6.7 | 6.5 | 26.5 | 68.9 |
| Canada | 7.2 | 7.5 | 23.5 | 70.9 |
| Germany | 7.9 | 8.3 | 28.4 | 65.9 |
| France | 8.6 | 9.2 | 32.0 | 62.0 |
| Finland | 9.1 | 8.9 | 25.4 | 67.7 |
| Italy | 9.5 | 9.2 | 39.3 | 54.9 |
| Spain | 10.7 | 11.2 | 34.2 | 58.8 |

Note: Unemployment based on OECD standards. These approximate to the ILO definition.

Source: OECD (2002), Tables A, B, F. Also Nickell (2003).

The case for including 'inactive' people in measures of unemployment is dependent on there being some similarities between the two groups in terms of their labour market aspirations and to a lesser extent the probability of these expectations being met. The DfEE recognised the need to help 'move people who are able to work off inactivity into the world of work', but viewed unemployed people and inactive people as two distinct groups (Education and Employment Committee, 2000).

However, the existence of a clear positive relationship across OECD countries between unemployment and inactivity suggests this assumption may not be correct: inactivity increases as unemployment increases (Figure 2.4). While not all of the inactive have aspirations to work, doubtless many of them would prefer and are able to do so, implying that the ILO definition of unemployment still seriously understates the real extent of worklessness.

One of the most disturbing aspects of the rise in inactivity is the steady growth in the numbers registered as unable to work because of sickness,

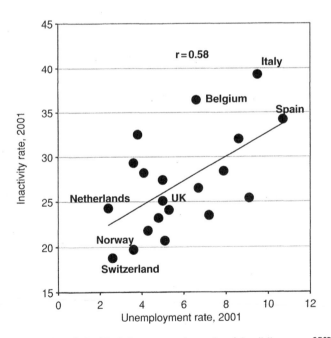

**Figure 2.4** The positive relationship between unemployment and inactivity across OECD countries. Source: Based on Table 2.1.

invalidity or incapacity. Again the UK case is striking. Up until the late 1970s, the numbers claiming some form of long-term sickness benefit averaged around 300,000–400,000. This number started to climb in the late 1970s, accelerated sharply in the 1980s, reaching 1.2 million by the end of that decade, and continued to grow throughout the 1990s, to stand at over 2.1 million in 2002 (Figure 2.5).

Several factors have contributed to this extraordinary trend. There is evidence that many of the claimants have come from the ranks of older male workers made redundant in heavy industries – such as coal mining and steel production – where health problems were unfortunately a common occurrence. The heavy deindustrialisation of these and similar traditional sectors in the 1980s left tens of thousands of male workers – many in local communities with few if any alternative employment possibilities – with two main options: to register for unemployment benefit, or to register for sickness and incapacity benefit. Since many of the older workers had little prospect of finding work, and were perhaps nearing retirement, the sickness benefit option was the more obvious one. High levels of unemployment were another factor: if the prospects of finding work are slim, and the individual can demonstrate some form of illness or incapacity that in any case limits the range of work he or she can do, then this increases the chances of being switched from unemployment to sickness support. At the same time, for many of the jobless, benefit levels for being 'on the sick' have actually exceeded those for being unemployed, thereby reinforcing the growth of the sickness total, and reducing the number officially classified as unemployed. So while there was unquestionably a genuine increase in the numbers of jobless who probably were unable to work because of sickness, the sheer size of the growth over the 1980s and 1990s was such that some at least of the individuals concerned should really have been classified as unemployed.

Thus while official unemployment in the UK has declined markedly since the early 1990s (especially when compared to other European countries), and despite recent claims by the Government for the success of the labour market policies introduced under the New Deal since 1998, this has not meant the end of the worklessness problem. For one thing, official claimant count and even Labour Force Survey measures of unemployment understate the extent of joblessness. For another, high levels of inactivity may be substituting, at least in part, for registered unemployment. And thirdly, as we now show, the worklessness issue, the hysteresis debate, and the interaction of unemployment and inactivity, all have an inherent geographical dimension that also confounds any simplistic claim that the unemployment problem 'has been solved'.

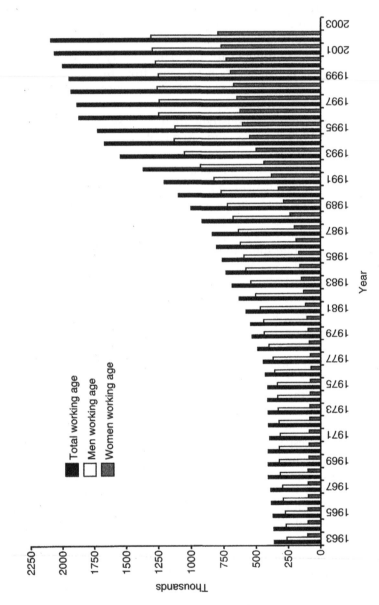

**Figure 2.5** Working-age claimants of sickness, invalidity and incapacity benefits (long term – over six months) UK. Source: Based on Webster (2003) and Office for National Statistics.

## The Geographies of Worklessness

In general, economists either ignore the geographical dimension of the labour market altogether, or if it is acknowledged at all, it is assumed to be relatively unimportant. The fact is that many labour market processes and outcomes have an intrinsic local level of operation (see Hanson and Pratt, 1992; Peck, 1996; Martin, 2001). There is no such thing as the 'national labour market', but rather a complex geographical mosaic of overlapping local and sub-national labour markets defined, and linked variously together, by a myriad of spaces of labour recruitment by firms and the job search, travel-to-work and migration fields on the part of workers. Although defining the geographical boundaries of local labour markets is fraught with difficulty – they are the dynamic product of processes of self-containment (integration) on the one hand, and of openness (interaction and interdependence) on the other – there is little doubt that geography acts to 'balkanise' or segment the labour market (Peck, 1996; Martin, 2001). Indeed, it is not just that local areas mediate or transmit wider macro-economic or even global forces impinging on employment and work, but that for many workers the very local constitution of the labour market imparts its own dynamic that intersects with these wider developments to shape and condition the particular opportunities and problems they face in particular places.

Just as there is no single all-embracing economic theory of the labour market, so geographers have debated a variety of theoretical perspectives on local labour markets and processes (Peck, 1996; Martin, 2001). At one extreme, neoclassical micro-economic models of the local labour market focus on the mechanisms of 'market-clearing', the supposed tendency for the local demand for and supply of different sorts of workers to equilibrate through wage flexibility and worker mobility between jobs and local areas. The failure for local labour markets to clear, as 'evidenced' for example by the presence of unemployment and vacancies, is taken to imply that 'barriers' and 'frictions' and 'impediments' are preventing the free and efficient allocation of labour. The theory is in effect less a theory of how local labour markets *actually* function, and more an ideological statement of how they *ought* to function in an ideal world.

In marked contrast to this model there are various heterodox representations that take local labour markets for what they are – imperfect, segmented by non-competing sub-markets, complex arenas in which employers and groups of workers – depending on local circumstances – exercise monopolistic or monopsonistic power, where historically evolved and locally embedded structures of employment, skills, and workplace relations become institutionalised and influence the job dynamics and wage structures of the area. Local, urban and regional labour market conditions and outcomes are thus likely to differ markedly. Because they differ in economic structure and institutional form, local labour markets operate in place-specific ways; they will also differ in their vulnerability to economic shocks and changes, as well as in their capacity to adjust to such developments. For example, the more economically specialised a local labour market is, the more traumatic will be the impact on local workers of negative shocks in the demand for labour (such as the closure of major employers and industries). The more so, because most workers are far less mobile than capital: the deindustrialisation of a community can take place almost overnight, but it takes time for labour to move to other areas: not only because of the constraints of housing, but also because of the pull of local social and family ties. To be sure, some groups of workers – typically the highly skilled and qualified, some of which are increasingly inter-national, even global, in their job search spaces – are more geographically mobile than others (such as the unskilled living in public housing). And workers in some countries are in general more mobile than those in others: geographical mobility in the US, for example, is reckoned to be three or even four times that in the UK (Table 2.2; see also Baddeley, Martin and Tyler, 2000). But the basic point is that it is in particular places that people live and work, and the conditions of that experience differ from place to place: geography matters.

Thus the rise in unemployment that characterised most OECD countries in the late 1970s and 1980s was far from uniform across regional and local labour markets within those economies. Geographical disparities in official unemployment rates widened markedly. Thus in the European Union, the regional dispersion (standard deviation) of unemployment rates rose from 2.0 in 1970 (when the EU-15 average unemployment rate was also 2.0 per

**Table 2.2**  Average net interregional migration in the UK and selected other countries compared (per cent of regional population)

| Period | US | Canada | Germany | Italy | UK |
|--------|------|--------|---------|-------|------|
| 1970–9 | 1.20 | 0.62 | 0.27 | 0.37 | 0.47 |
| 1980–9 | 0.84 | 0.63 | 0.34 | 0.33 | 0.26 |
| 1990–5 | 0.87 | 0.52 | 0.31 | 0.40 | 0.20 |

Note: Figures are population-weighted averages over regions. For the period concerned, each regional figure is calculated as the average absolute value of the change in regional working age population (measured net of national working-age population growth). German figures are for western Länder only, excluding Berlin.

Source: Obstfeld and Peri (2004).

cent) to 3.0 in 1980 (when the EU-15 unemployment rate was 6.0 per cent) to 5.0 in 1984 (EU-15 rate of 10.6 per cent) to 6.0 in 1994 (EU-15 rate of 11.4 per cent) (see also Martin, 1997). Although unemployment has since declined in the EU as a whole, regional disparities in many Member States remain marked (see Table 2.3).

The same has been true of the UK. Regional unemployment rate disparities increased sharply as national unemployment rose during the 1980s. By 1986 they had reached their widest for more than 50 years, and mapped out a distinctive pattern that was widely referred to as the 'north–south divide' (see Martin, 1988, 1993, 2004). In fact, as Figure 2.6 shows, regional disparities in joblessness mapped out a threefold division: a southern 'core' with unemployment below 9 per cent (south-east, London, East, South-West and East Midlands); a 'semi-periphery' with claimant rates of around 12 per cent (Yorkshire-Humberside, West Midlands, Scotland, Wales, and the north-west); and a 'periphery' with rates of 15 per cent or

**Table 2.3**   Regional extremes in unemployment across EU-15 member states, 2002 (NUTS2 regions)

|  | National Average Unemployment rate | Highest Unemployment Rate Region | Lowest Unemployment Rate Region |
|---|---|---|---|
| Austria | 4.0 | 7.2 | 2.0 |
| Belgium | 7.5 | 14.5 | 3.8 |
| Denmark | 4.6 | na | na |
| Finland | 9.1 | 14.1 | 2.9 |
| France | 8.7 | 13.4 | 6.4 |
| Germany | 9.4 | 27.1 | 3.8 |
| Greece | 10.0 | 14.7 | 7.3 |
| Italy | 9.0 | 24.6 | 3.3 |
| Ireland | 4.3 | 5.5 | 3.8 |
| Luxembourg | 2.6 | na | na |
| Netherlands | 2.8 | 4.2 | 2.2 |
| Portugal | 5.1 | 6.6 | 2.5 |
| Spain | 11.4 | 19.2 | 5.3 |
| Sweden | 5.1 | 6.3 | 3.9 |
| United Kingdom | 5.1 | 9.0 | 3.5 |

Notes: na – not applicable (no NUTS2 regions)
France excludes overseas departments
ILO measure of unemployment.

Source: Eurostat.

more (north-east and Northern Ireland). During the rapid economic boom of the second half of the 1980s, as national unemployment fell, so regional disparities narrowed (with the exception of Northern Ireland). Then, in contrast to the recession of the early 1980s, in the downturn of the early 1990s regional unemployment disparities continued to narrow rather than widening once more. Whereas the earlier recession had severely impacted on the manufacturing sectors of the northern regions of the country, that of a decade later was much more centred on the service economy of southern Britain. London in particular experienced a marked rise in unemployment, and for a while became the third highest unemployment region (Figure 2.6). As economic recovery progressed during the 1990s, so regional disparities continued to narrow, so that by 2000 they were almost back to the level of the mid-1970s.

Perhaps not surprisingly, therefore, by the late 1990s talk of a 'north–south' jobs divide had all but ceased, and some observers were proclaiming the end of Britain's 'regional unemployment problem':

> The traditional 'North-South' unemployment problem has all but disappeared in the 1990s. This may prove to be a permanent development since the manufacturing and production sector, the main source of regional imbalance in the past, no longer dominates shifts in the employment structure to the same extent. Future shocks will have a more balanced regional incidence than has been the case in the past (Jackman and Savouri, 1999, p. 27).

But while regional unemployment differences have narrowed, local disparities within regions have not only persisted but have increased. As Green, Gregg and Wadsworth (1998) showed, as early as 1993, within-region disparities had become much more marked than between-region disparities, whereas previously both sources of variation had been roughly the same magnitude. In fact, local unemployment rates have actually become more dissimilar (see Figure 2.7). By the end of the 1990s, local unemployment rates varied by a factor of seven, from a low of 2.0 per cent to a high of 14.0 per cent (Table 2.4). As Table 2.4 and Figure 2.8 show, high unemployment areas tend to be of three kinds: older industrial and urban areas in the Midlands and northern regions of the country (such as Birmingham, Sunderland, Manchester, Liverpool, Newcastle, Middlesborough, Glasgow); rural areas and coastal towns (such as much of West and South Wales, the highlands of Scotland, Thanet in the south-east region, and Great Yarmouth in the east of England); and a number of boroughs in inner London (such as Hackney, Tower Hamlets, Newham, Islington). However, although local variations are found in every major region of the country, most of the low unemployment districts continue to be found in

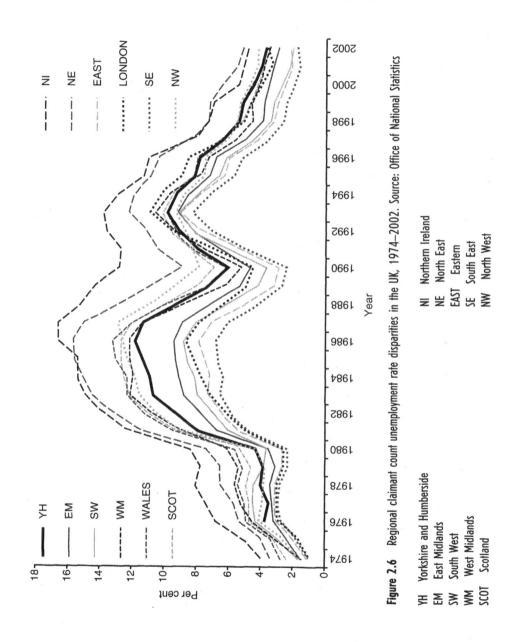

**Figure 2.6** Regional claimant count unemployment rate disparities in the UK, 1974–2002. Source: Office of National Statistics

YH   Yorkshire and Humberside            NI     Northern Ireland
EM   East Midlands                       NE     North East
SW   South West                          EAST   Eastern
WM   West Midlands                       SE     South East
SCOT Scotland                            NW     North West

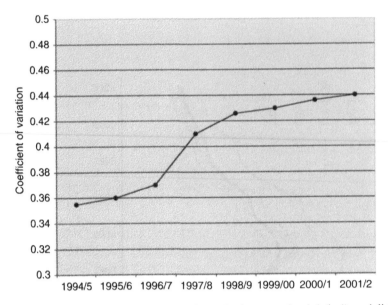

**Figure 2.7** Increasing local variation in unemployment rates across Local Authority and Unitary Authority districts in the UK, (ILO based Unemployment Rules), 1994/5 to 2001/2. Source: Office for National Statistics.

southern England, especially in the south-east region outside London. This picture is reinforced by the local variation in the working-age employment rate (Figure 2.9).

Thus even though the second half of the 1990s was a period of increasing buoyancy in the overall UK labour market, and national official unemployment fell accordingly, it is clear that many local areas failed to benefit fully from this improvement, and in effect got left behind. Most of these high-unemployment local labour markets are areas that have suffered a major erosion of their economies – through deindustrialisation, run-down of former defence sectors, loss of major employers, urban degradation, and so on – and have been slow to generate alternative sources of jobs. The spatial connection between job loss and unemployment has been spelled out for cities by Turok and Edge (1999), for the coalfields by Beatty et al. (1997), and at regional level by Martin (1993, 1997) and Rowthorn (2000). As mentioned above, economic adjustment and revival is particularly difficult at the local level. And for this reason, it is perhaps at the spatial level that the issue of hysteresis does take on meaning and significance. Local areas, once hit by severe economic depression, can become caught in a situation where both demand-side and supply-side processes interact to maintain high unemployment locally even though employment conditions in the economy at large have improved.

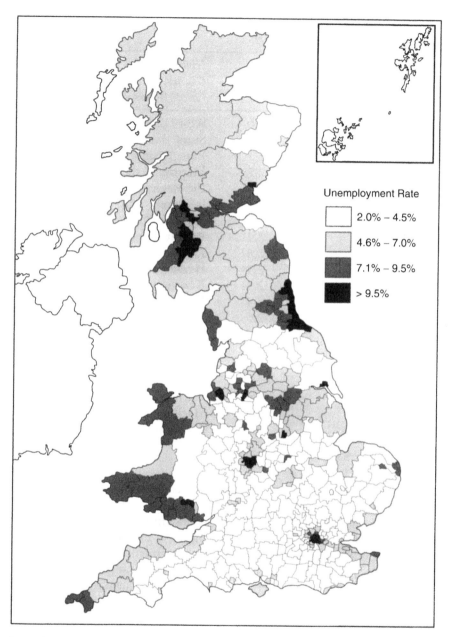

**Figure 2.8**  Local ILO unemployment rates (by Local and Unitary Authority areas) across the UK, 1999–2000. Source: Office for National Statistics.

**Table 2.4** Twenty lowest and highest unemployment localities in the UK at the end of the 1990s (ILO-based unemployment rates, 1999–2000)

| Lowest Unemployment Local Authorities | Rate | Highest Unemployment Local Authorities | Rate |
|---|---|---|---|
| Hart | 2.0 | West Dunbartonshire | 10.4 |
| Surrey Heath | 2.0 | Sunderland | 10.4 |
| West Oxfordshire | 2.1 | East Ayrshire | 10.5 |
| Woking | 2.2 | Merthyr Tydfil | 10.6 |
| South Northamptonshire | 2.3 | Birmingham | 10.9 |
| Wokingham | 2.3 | Islington | 10.9 |
| Vale of White Horse | 2.4 | Dundee City | 11.0 |
| West Berkshire | 2.4 | Redcar and Cleveland | 11.0 |
| Mole Valley | 2.4 | Liverpool | 11.2 |
| Cotswold | 2.4 | Haringey | 11.2 |
| Waverley | 2.4 | Knowsley | 11.3 |
| Cherwell | 2.4 | Newcastle upon Tyne | 11.6 |
| North Dorset | 2.5 | South Tyneside | 11.7 |
| Bracknell Forest | 2.5 | Manchester | 11.7 |
| Uttlesford | 2.5 | Newham | 12.4 |
| Ribble Valley | 2.6 | Hartlepool | 12.5 |
| South Cambridgeshire | 2.6 | Tower Hamlets | 12.5 |
| Test Valley | 2.6 | Glasgow City | 12.7 |
| Forest Heath | 2.6 | Hackney | 13.0 |
| Wealden | 2.6 | Middlesbrough | 14.0 |

Source: Office for National Statistics (2003).

On the demand side, the loss of large sections of the local economic base, high unemployment, and low incomes, all make the locality unattractive to new investment and business, and hence employment growth remains low. The local variation in employment rates shown in Figure 2.9 testify to the geographically uneven nature of employment growth across the UK in recent years. Job growth has been primarily concentrated in southern Britain, although even there certain local areas have failed to share in the process. Conversely many localities in northern Britain have recorded little if any net employment growth, and some have even experienced falls in employment.

At the local level the lack of labour demand creates and reinforces problems on the supply side. For what is also particularly striking about these local disparities in unemployment is the way in which they correlate closely with the problems of long-term joblessness and economic inactivity.

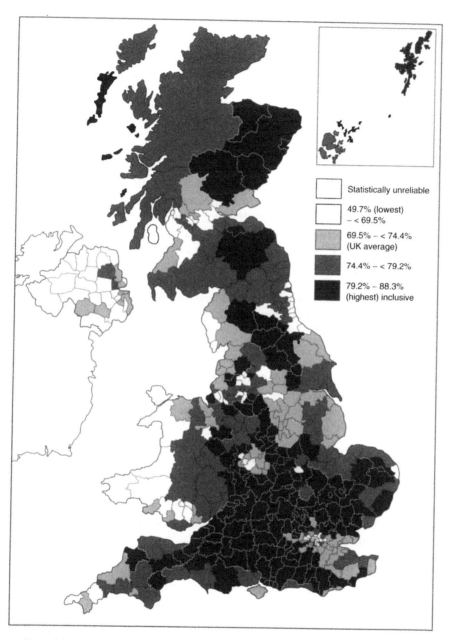

**Figure 2.9** Local Working Age Employment Rates (by Local and Unitary Authorities) across the UK, 2001–2. Source: Labour Force Survey, Office for National Statistics.

Legend:
- Statistically unreliable
- 49.7% (lowest) – < 69.5%
- 69.5% – < 74.4% (UK average)
- 74.4% – < 79.2%
- 79.2% – 88.3% (highest) inclusive

The higher the local rate of unemployment, the higher also the rate of long term unemployment (see Figure 2.10). In addition, as both unemployment and long-term unemployment increase across local areas, so do both the total inactivity rate and the proportion of the local population claiming sickness and disability benefits. This implies three things.

First, even though local disparities in official unemployment at the end of the 1990s were substantial, these seriously understated 'real' unemployment differences, since there were also large local disparities in economic inactivity, and hence in hidden worklessness. Local inactivity rates in early 2001 ranged from 6 per cent to 34 per cent. While not all inactivity represents hidden unemployment of course (e.g. the high rates in the university towns of Oxford, Cambridge and Durham reflect the large student populations there), nevertheless in many localities it is likely that a significant proportion of the inactive would like to work if local employment opportunities were more plentiful and attractive. Second, the relationships in Figure 2.10 also suggest that this understatement effect is likely to be most serious in the localities that already have the highest official unemployment rates. In the areas that have the worst unemployment and the highest rates of long-term unemployment, a higher proportion of the working age population drop out of the local labour market altogether, into inactivity and on to sickness benefits. As a consequence, the real unemployment rate in such areas is almost certain to be considerably higher than official statistics indicate. And third, these processes suggest that high local unemployment gives rise to various supply-side effects that tend to reproduce high unemployment, long-term unemployment and economic inactivity: in other words, a sort of localized hysteresis effect (see Gordon, 2003).

Some of the possible processes involved are shown schematically in Figure 2.11. For example, in a depressed local labour market many workers are likely to experience shorter employment spells. This can erode or militate against the development of skills, and perhaps even undermine workers' job-search motivation. An individual's occupational attainment is hindered thereby, making that person less able to find permanent stable employment, and hence more prone to unemployment. Or again, high local unemployment can disrupt family relations and structures, leading to family fragmentation and a whole host of social problems, including the educational underachievement of young people. Members of such families are known to find it difficult to secure and hold on to jobs, particularly under slack local labour demand conditions, and where the demands of single parenting clash with regular working hours. Yet further, repeated spells of unemployment, or long periods of joblessness, increase the chances of ill-health, which in turn reduces the employment prospects of the individual

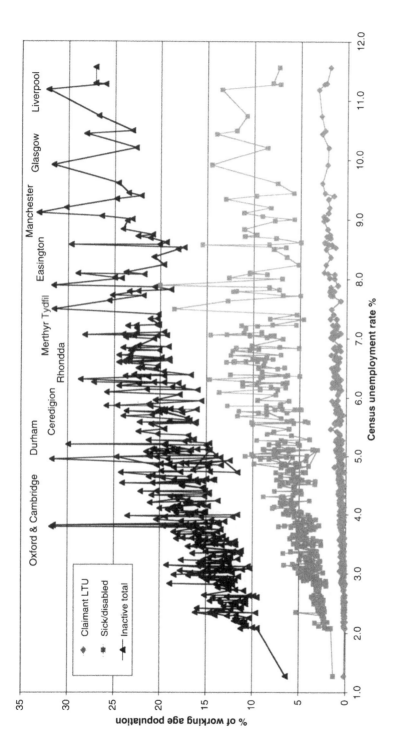

**Figure 2.10** Local (UA/LAD) rates of long-term claimant unemployment, sick/disabled, and economically inactive, by census unemployment rate, males, April 2001.
Source: Adapted from Webster (2003).

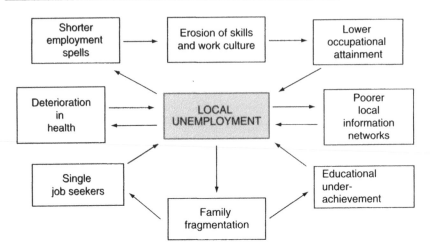

**Figure 2.11**   Processes of hysteresis in high-unemployment localities. Source: Based on Gordon (2003), Figure 3.3 Causal links in the reproduction of concentrated employment, page 77 in R. Martin and P. Morrison (eds) *Geographies of Labour Market Inequality*, Routledge.

concerned, and increases his or her chances of unemployment or even withdrawal from the labour market into inactivity. And the presence of high levels of unemployment tends to destroy and hinder the development of local information networks about jobs and employment opportunities: the unemployed become disconnected and excluded from such networks, and thus more likely to remain unemployed.

It is important to be clear about what is being suggested here. It is not being argued that high unemployment rate localities owe their worklessness and long-term unemployment problems simply or solely to the particular characteristics of their local workforce or labour supplies. Rather, the contention is that when demand for labour shifts or collapses in a locality, the resultant high unemployment can have negative effects on the characteristics and 'employability' of many of those unemployed that in turn diminish their chances of getting back into work and hence increase their chances of becoming repeatedly, or long-term, unemployed. Under these conditions, it becomes particularly difficult for young labour force entrants, especially those with few skills or low levels of educational attainment, to gain a foothold in work. Supply-side problems can thus be the result of demand-side problems. This point is critical, for it has major implications for the design, purpose and impact of labour market policies targeted and implemented at the local level.

When the UK New Deal programme was introduced in 1998, it was focused firmly on the young unemployed, the 18–24 year old group of jobless. Although the same basic welfare-to-work approach has since been extended to several other disadvantaged groups in the labour market, the New Deal for Young People (NDYP) has remained the flagship policy. The problem of youth unemployment was viewed as particularly disturbing since sizable numbers of 18–24 year olds out of work raised the prospect of large sections of this part of the nation's workforce becoming permanently excluded from the labour market, and from the socio-economy more generally. Young people with low levels of educational attainment were seen as especially prone to unemployment upon entering the labour force, particularly given the shifting contours of employment towards more skilled and qualified workers, noted at the beginning of this chapter. Others argued, however, that the problem had more to do with the fact that wages for young people had become too high relative to older, more skilled workers, that in effect many young people had become 'priced out of work'. To this others argued that overly generous and unconditional unemployment and income benefits merely compounded the problem and discouraged young workers from job search and employment.

Absent from these discussions and debates, as in the case of unemployment more generally, was any serious discussion of the geography of youth worklessness. The causes of youth unemployment were seen to be national and generic in scope, and not to have any locally specific dimension. Yet, as with total unemployment, youth joblessness varies considerable across the country.

In 2003, for example, local claimant unemployment rates for the 18–24 year old group varied by more than a factor of 12, from 7.6 per cent to 0.6 per cent (Table 2.5). While all parts of the UK contain local areas of high and low youth unemployment, the overall geographical pattern tends to follow that of total unemployment. High rates of youth unemployment tend to characterise the inner areas of many of Britain's northern cities and conurbations (such as Hartlepool, Middlesborough, Liverpool, Newcastle, Tyneside, Teeside, Birmingham, Glasgow and Manchester), and in the problem boroughs of inner London (Tower Hamlets, Lambeth, Hackney, Southwark, Lewisham and Islington). In some of these localities, jobless young people account for 30 per cent or more of the total unemployed (e.g. Oldham, Knowsley, Sunderland, Teeside, Bolton, Wigan, Manchester and Liverpool). Conversely, the lowest youth unemployment rates tend to be concentrated in southern Britain (typified by Basingstoke, Cambridge, Guildford, Reigate, Winchester, Wokingham, Mid-Sussex, and Runnymede). In many of these areas, the young make up only around 15 per cent or less of the total unemployed. Local youth inactivity rates also vary considerably across the country – from 13 per cent to 55 per cent, but the

**Table 2.5**  Twenty lowest and highest youth unemployment localities in the UK, 2003 (average annual claimant count rates[*])

| Lowest Youth Unemployment Local Authorities | Rate | Highest Youth Unemployment Local Authorities | Rate |
| --- | --- | --- | --- |
| Hart | 0.60 | Tower Hamlets | 5.93 |
| Mole Valley | 0.60 | Blaenau Gwent | 5.94 |
| Runnymede | 0.63 | Wear Valley | 5.96 |
| Rutland | 0.68 | Cleveland | 5.98 |
| Waverley | 0.69 | Wolverhampton | 6.05 |
| Mid Sussex | 0.79 | Merthyr Tydfil | 6.06 |
| Uttlesford | 0.80 | Sandwell | 6.13 |
| West Oxfordshire | 0.82 | Lewisham | 6.21 |
| Wokingham | 0.84 | South Tyneside | 6.22 |
| Winchester | 0.86 | Southwark | 6.36 |
| Tandridge | 0.86 | Lambeth | 6.52 |
| Ribble Valley | 0.87 | Hackney | 6.56 |
| Vale of Whiter Horse | 0.87 | East Ayrshire | 6.65 |
| Surrey Heath | 0.89 | Kingston upon Hull | 6.65 |
| Reigate | 0.89 | Wansbeck | 6.82 |
| East Dorset | 0.91 | Knowsley | 6.87 |
| South Northamptonshire | 0.91 | Middlesborough | 7.15 |
| North Dorset | 0.92 | North Ayrshire | 7.28 |
| Guildford | 0.93 | Inverclyde | 7.35 |
| South Cambridgeshire | 0.94 | Hartlepool | 7.65 |

[*]These rates are probabe underestimates as we have only been able to obtain local population figures for the 15–24 age group, whereas the unemployment count is for 18–24 year olds.

Source: Office for National Statistics (2003).

geography of these is more difficult to interpret, since the figures include young people in further education.

The main point is that youth unemployment, like total unemployment (of which, of course, it is a part) is highly uneven in its geographical incidence. Thus, even in 2003 – after several years of operation of the New Deal for Young People – youth unemployment still exhibited different degrees of concentration across the country, relative to both the local youth labour force and the local total number of unemployed. As in the case of total unemployment or long-term unemployment, these disparate geographies cannot be explained in purely supply-side terms, as deriving from the characteristics and attitudes of individuals, but have much to do with local variations in labour demand, and how that demand intersects with labour supply. The message is clear: local labour market conditions matter.

## The Policy Challenge

From its inception in 1998, the UK New Deal programme for the unemployed has focused on the supply-side: the aim of the policies under the programme has been to improve the 'employability' of the unemployed so as to make them attractive to employers and more active in their job search and motivation to work. The above discussion suggests that there is indeed a vital role for such policies (which were conspicuous by their absence for much of the post-war period). But a belief that somehow 'an improved supply of labour will create its own demand' – a sort of crude Say's Law applied to the unemployed – is problematic. If insufficient numbers of suitable jobs are available locally, there is no guarantee that those newly entering the labour market and those coming off New Deal training, work experience and further education programmes will find work. Indeed, the spatialities of unemployment and inactivity pose a key paradox for locally delivered, supply-side orientated policies. On the one hand, supply-side measures need to be delivered on a local basis, to respond to the specific personal characteristics and skill problems of the locally unemployed. But on the other, by focusing only on the supply-side, and ignoring local demand-side conditions, such policies run the risk of having the least impact where they are most needed: in those local areas with insufficient employment opportunities due to a lack of local demand for labour. In other words, supply-side policies, even if locally based and administered, could end up having the least impact in those localities with the worst unemployment problems.

In terms of its very design, the New Deal programme has devoted little attention to the demand side. Essentially, it has been argued that local variations in labour demand have been either insignificant or unimportant. Indeed, Richard Layard, one of the academic exponents of New Deal type policies in the UK, has repeatedly claimed that local disparities in unemployment – to the extent that they exist – should have no detrimental effect on the impact of the policy. And the Government itself has maintained that there is no real problem of local variations in labour demand, in that plenty of vacancies exist everywhere, or are certainly within easy geographical reach of the locally unemployed.

However, our analysis in this chapter suggests that local variations in unemployment across the UK have been large and apparently entrenched. It is difficult to argue that these differences are solely or even primarily attributable to local differences in the supply-side of the labour market. The evidence points instead to significant and enduring local differences in labour demand conditions. To be sure, as we have discussed above, these local differences in demand may feed back to influence local supply-side conditions, thereby exacerbating local differences in unemployment. But to the extent that demand conditions have indeed differed across local labour markets in the UK, a policy orientated predominantly to supply-side interventions, as the New Deal has been, is likely to have locally variable success. It is thus to an assessment of the impact of the original New Deal programme (for young people) on the UK's local geographies of unemployment that we now turn.

# Chapter Three

# Local Disparities in the Performance of Welfare-to-Work

## Introduction: Residual Pockets and the Geography of New Deal

The Labour Government has repeatedly described its welfare-to-work policies as one of its biggest successes. It is claimed that they have helped to deliver record levels of employment and 'one of the world's strongest labour markets' (Smith, 2004, p. 1). Unemployment, it is noted, has fallen to its lowest level for 30 years, and the employment rate is the highest of the seven major industrialised countries. The New Deal programmes, and especially the New Deal for Young People (NDYP), have been presented as being catalysts to this success. The NDYP has been self-funding, and by 2001 long-term youth unemployment had – in the view of the Secretary of State for Employment – been 'virtually eliminated' (Department for Education and Employment [DfEE], 2001a).[1] In 2004 the Government announced that over 493,000 unemployed young people had been moved into work and that without the policy youth unemployment would be twice as high (Department for Work and Pensions, 2004a, p. 2). It is claimed that the UK has become a 'world leader' in welfare-to-work policies and that other countries can learn from its success (HM Treasury, 2003). According to the Minister for Work in 2004, 'We are at the leading edge of labour market policies and representatives from all over the world come to see Job Centre Plus and the New Deal programmes in action' (Browne, 2004a, p. 1).

This official judgement has contained no hint that the NDYP may have more successful in some places than others. Initially there was little willingness on the part of Government to acknowledge that there had been any variability in the programme's outcomes in different parts of the country. In the context of Labour's sensitivity about the relative economic standing of its traditional northern manufacturing electoral heartlands, comments about spatial differences in the outcomes of the New Deal welfare-to-work programme were firmly 'off-message'. In addition, at a relatively early stage in the development of the New Deal, the question of geography became strongly associated with a politically unwelcome demand-side critique of the programme. Critics such as Turok and Webster (1998) and Peck (1999) argued that the concentrated geography of unemployment would undermine the programme because the job gap in a depressed labour market severely limits the chances of successful permanent work placement of individuals processed by the scheme. Thus, any political acknowledgement that the impact of the New Deal has been variable across regions and cities seemed to risk conceding some ground to this critique, and thus appearing as an admission that there was after all a resilient 'jobs-gap' in some parts of the country. According to the Employment Minister Tessa Jowell, 'the figures show that New Deal is working in every region of the UK...It proves that New Deal is able to meet the needs of different regional labour markets and different sectors of the economy' (DfEE, 2000, p. 1). A short Employment Service (2000) report claimed that 'Places all over Britain have benefited from the New Deal. Hardly a town or a city or area throughout the country have not seen youth unemployment fall, and the gains from the New Deal appear to be as strong in the North, Scotland and Wales as in the South of England'(p. 18). This conclusion, however, was based solely on percentage falls in long-term youth unemployment in a small selection of Local Authority Districts.

This overwhelmingly positive and celebratory official view has, nevertheless, recognised the need for a certain degree of policy reform and improvement. While most evaluations suggest that the New Deal has broadly met its targets, it has been noted that the programme has been marred by high and unsatisfactory levels of *recycling or churning* of individuals between work and unemployment. For example, the Education and Employment Committee's (2001a) *Report on the New Deal* expressed concern that the proportion of moves into unsustained employment (defined as jobs lasting less than 13 weeks) remained as high as 40 per cent. It also noted that over 20 per cent of participants were re-entrants who failed to get employment the first time round, and recommended that job retention and career progression should be built into future assessments of the New Deal.

The dominant way in which the appearance of such workfare recycling has been explained is more debatable. It is typically argued that the success of the programme in a context of a favourable national economic climate has exposed the most unemployable individuals who suffer from multiple and acute personal obstacles to work. Thus it is argued that some reforms are needed to provide additional support to this *'hard core' of unemployable individuals*, such as the fuller development of basic literacy, numeracy and core work skills in the Gateway stage of the programme, or a broader range of training options. While this view is undoubtedly more applicable in some local labour markets than others, it is immediately striking that this interpretation is entirely consistent with the paradigm of supply-side economic thought that framed the New Deal in the first place (see Chapter 2).

However, recent policy documents also admit that some aspects of this 'hard core' unemployability problem are geographical. They suggest that in a buoyant national labour market, worklessness is highly concentrated in certain groups of the population and in local wards and districts. As we noted in Chapter 1, the recognition of geographical variability is limited to the claim that there are 'residual pockets' of worklessness in localised areas. Thus HM Treasury and DWP argue that employment has risen and unemployment fallen in every region (with the gains in employment being greatest in regions with weakest starting positions) but that 'the benefits of this improvement have not been felt by all groups or across all areas to the same degree, and, at a local level, there remain severe concentrations of worklessness' (2003, para 1.15; see also DWP, 2004a). They later add that 'the worst concentrations of joblessness are in very small defined areas and *are not caused by a lack of jobs*, but by the people living in these areas being unable to compete successfully for the vacancies available'(2003, para 4.32, emphasis added).[2] Thus even in these areas where it is acknowledged that a severe joblessness problem remains, and in stark contrast to the arguments of critics like Turok, Peck and others, the cause of such localised unemployment is still interpreted as due not to a lack of labour demand but to inadequacies of labour supply. It is argued that national policies alone will not be enough to address concentrations of disadvantage as 'in some areas, "a culture of worklessness" or "poverty of aspirations" has developed, locking people into cycles of worklessness' (HM Treasury and DWP, 2003, para 4.34). In all cities a large number of people claiming unemployment and inactivity benefits coexists with a high number of jobs and vacancies, and there are more jobs than residents demonstrating that in-commuters can often better compete for available vacancies. Thus the problem is defined in terms of the failure of the unemployed to connect with the jobs that are available, a failure of aspirations rather than one of opportunities.

This chapter begins to evaluate how far this 'residual pockets' and 'localised concentrations of disadvantage' argument adequately describes the geographical impact of the NDYP by examining the extent of local and regional variations in its performance. While there have been several local case studies of the outcomes of the NDYP (e.g. see Hoogvelt and France, 2000; Gray, 2001; Hyland and Musson, 2001), there have been relatively few systematic and extensive studies of how the programme has had varying impacts in different types of local labour market across the country as a whole. The chapter incorporates some of our previously published analysis of the geography of the programme's effects during the first two years of its operation when, in aggregate terms, it was having its strongest impact on the problem of youth unemployment (see Sunley, Martin and Nativel, 2001). We then extend these findings with an examination of some of the geography of programme results in the period since 2000. Unfortunately, this is far from straightforward, given the change in the official presentation of New Deal performance statistics during 2000, which complicates comparison before and after that date. Prior to 2000 a wide range of core performance indicators were published for the local programme areas – entitled 'units of delivery' (UoDs). Since mid-2000 a much reduced set of indicators has been made available for larger Job Centre Plus Districts (see the appendix at the end of this chapter). While we have managed to obtain and compile some consistent data series, as a result of this statistical disjuncture much of the analysis that follows is forced to distinguish the first two years of the programme from the period since 2000.

The chapter examines geographical variations in some of the New Deal's key outcomes at the level of the UoDs – through which the scheme was implemented until 2000 (for a discussion of the nature of these areas, how they relate to 'local labour markets', and how they were succeeded by Job Centre Plus Districts from 2000, see the appendix). The first part of the chapter uses the programme's own 'core performance' measures to look at its varying local effectiveness. These recorded how each UoD was meeting key aims such as moving young people into subsidised and unsubsidised jobs, and the proportion of these remaining in jobs for 13 weeks. Until 2000 these outcomes were frequently expressed as a proportion of the cohort of clients entering the programme during a specified three-month period. On the basis of these entry cohort statistics, and more recent indicators, we argue that the neglect and marginalisation of geography in official evaluations is regrettable, as the spatial variations in the policy's performance have in fact been significant. Partly as a consequence, geographical variations in the relative severity of youth unemployment show little sign of convergence across different local labour markets.

The second part of the chapter turns to the relation of labour market flows to the programme's outcomes. Once again, most discussions of the New Deal's effects on flows of young people into and out of claimant unemployment have been aggregate and national-scale (see Anderton et al., 1999; Riley and Young, 2000). There is little published work on whether there have been geographical variations in inflows and outflows to youth claimant unemployment during the period of the programme's operation. The chapter therefore seeks to provide such evidence. While data constraints prevent us from distinguishing the local-level changes in flow numbers and rates specifically attributable to the New Deal, our analysis does nevertheless show that inflows to claimant unemployment during the first two years of the programme's operation behaved in strikingly different ways in different UoDs.

The third section of the chapter considers whether our argument about the significance of local space and context applies with equal force to other welfare-to-work programmes. It maps some of the local results of the New Deal (ND) for 25 Plus and reviews some of the literature on the achievements of the New Deal for Lone Parents, and considers how far local context has also affected these programmes. Both of these programmes have to some degree been reformed in an effort to copy some of the key features of the New Deal for Young People (NDYP) and to replicate its success. But what is also clear is that these programmes, and especially the ND 25 Plus, are working much more effectively in some local labour markets than others, and are also showing strong geographical disparities in their outcomes.

## Mapping New Deal Outcomes

While much about the local dimension of the NDYP is uncertain, it is not controversial to say that at its inception in April 1998 the programme faced a widespread but highly locally uneven problem of unemployment. As we have seen in the previous chapter, the spatial incidence of unemployment in the late 1990s was far from spatially even (see also Turok and Webster, 1998; Green and Owen, 1998). While the geography of youth unemployment showed a broad regional 'north–south divide' (see Martin et al., 2001), long-term youth unemployment was particularly concentrated in the inner areas of the conurbations and large cities, including London. In the first quarter of 1997 a mere 10 out of 144 New Deal UoDs (Birmingham, Manchester, the Black Country, Liverpool, Glasgow, Newcastle, Tees North, Sheffield, Edgware and Hackney), or less than 1 per cent, accounted for some 22 per cent of the total number of unemployed 18–24 year olds. Most rural UoDs faced the profoundly different task of dealing with smaller totals of, and often more spatially dispersed, long-term youth unemployment. Figure 3.1 shows the claimant count unemployment rate for those aged 18–24 years who had been unemployed for six months or more, averaged across 1997, as a percentage of the total population in this age group, for 90 or so UoDs.[3] In broad terms this indicates that long-term youth unemployment rates varied across the highest and lowest UoDs by a factor of nine or ten. Figure 3.2 maps the incidence of youth unemployment as a proportion of total unemployment for 1997 for all 144 UoDs, and reveals that the severity of the problem to be tackled was much greater in northern conurbations and in industrial regions than in most of the south of the country. Contrary perhaps to the impression given by official statements on the New Deal at the time, it was clearly introduced in a context of wide local variations in the very problem it was intended to address: indeed, as noted in the previous chapter, the evidence is that by the late 1990s local unemployment variations across the UK had become considerably greater than regional differences (also Green, Gregg and Wadsworth, 1998). The key question is whether and to what extent such local variations in labour market conditions have themselves influenced the impact of the NDYP.

Our assessment of this issue begins by mapping some of the New Deal 'core performance measures' made available by the Employment Service for the first two years or so of the NDYP's operation. Evaluating the early performance of the programme in terms of its own priorities, Figure 3.3 shows the average percentage of programme participants in the first seven quarterly cohorts (those who joined the NDYP between April 1998 and December 1999) who had obtained unsubsidised jobs by April 2000. In many ways this was seen as the most important official performance measure, since the immediate policy priority is to get people into jobs, on the assumption that employability will follow. While the results show quite a complex pattern, reflecting the combined effects of local labour market conditions and local New Deal office management efficiencies (see Nativel, Sunley and Martin, 2003), it is possible to draw some conclusions (for a description of the different local management delivery models, see Chapter 6 and the appendix). It is clear that on this measure, the NDYP was most effective in many rural UoDs, where on average 50 per cent and over of programme participants were moved into jobs. This is particularly the case in Southern England, but much less so in Wales and Scotland. We should add that those local areas in northern Britain with high rates of job placement (particularly in North Yorkshire, East Lancashire, West Lothian and Cheshire) are precisely those that did not face a particularly severe youth unemployment problem to begin with (compare Figures 3.2 and 3.3). In contrast, all of the major inner conurbations (with the exception of Manchester) have the lowest rates of job placement.

However, the picture is changed in some ways if we add subsidised job placements into the evaluation (as Chapter 2 noted, this was one of the four options available to NDYP participants). Figure 3.4 shows the proportion of the same cohorts who had by April 2000 found either unsubsidised jobs or had entered the subsidised employment option. Some rural UoDs in Wales, Scotland, Cumberland and the West Country appear to have had more success in finding subsidised job placements. But in much of the Midlands and Central Southern England the jobs subsidy has made less of an impact on job outcomes. What emerges is the relatively poor results for the major conurbations and larger cities such as Sheffield and Nottingham.

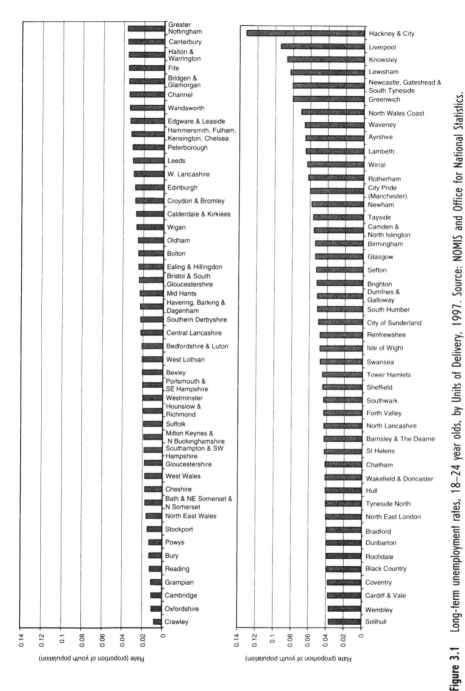

**Figure 3.1** Long-term unemployment rates, 18–24 year olds, by Units of Delivery, 1997. Source: NOMIS and Office for National Statistics.

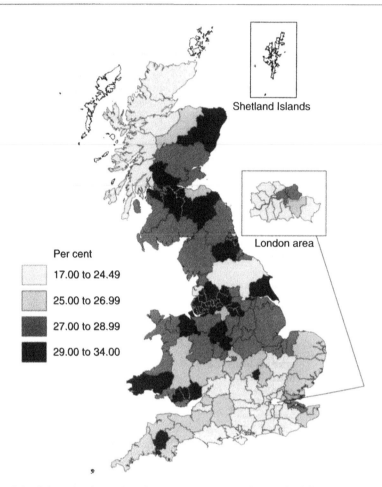

**Figure 3.2**    Relative incidence of youth unemployment across the New Deal Units of Delivery areas (annual average for 1997). Source: NOMIS and Office for National Statistics.

The performance of the programme since 2000 shows a similar pattern of geographical disparity. Figure 3.5 shows the job entry rate for all UoDs up until March 2002. The lowest rates are in the conurbations, in the industrial North and Midlands and in areas around London. Again the relatively poor performance of much of central England and Scotland is evident. Figures 3.6 and 3.7 map job attainment rates for Job Centre Plus Districts for the period between January 1999 and June 2003. Despite the change in boundaries, the same broad patterns are evident, although the poor performance of London is even more marked for this period. Figure 3.7 shows the attainment of unsubsidised sustained (13-week or

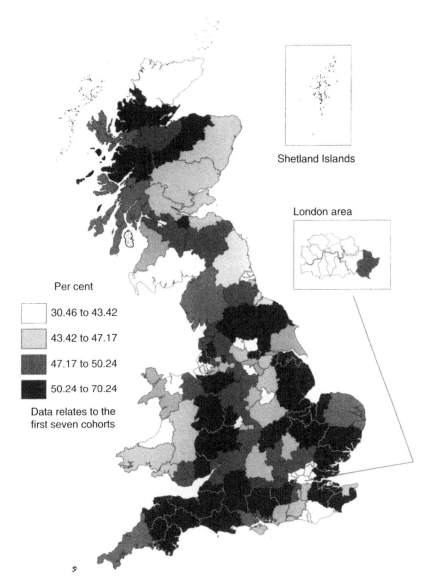

Per cent

30.46 to 43.42

43.42 to 47.17

47.17 to 50.24

50.24 to 70.24

Data relates to the
first seven cohorts

Shetland Islands

London area

**Figure 3.3** Proportion of New Deal participants entering unsubsidised jobs, April 2000. Source: New Deal indicators, Department for Work and Pensions.

more) jobs and reveals a strong urban–rural contrast with the highest rates in rural districts in the North, in the East, South-West, mid-Wales and the Highlands. It further confirms the less successful performance of the programme in the core cities and in London.

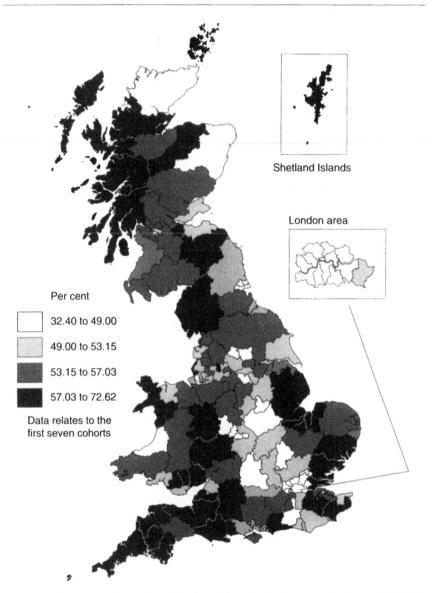

Per cent

32.40 to 49.00

49.00 to 53.15

53.15 to 57.03

57.03 to 72.62

Data relates to the
first seven cohorts

Shetland Islands

London area

**Figure 3.4**  Proportion of New Deal participants entering subsidised or unsubsidised jobs, April 2000.
Source: New Deal indicators, Department for Work and Pensions.

Further evidence for this is provided if we examine the proportion of
entrants finding jobs between 1998 and 2000 across the different types of
UoD 'cluster' as defined by the Employment Service. In order to compare
the performance and management of similar types of local areas, the

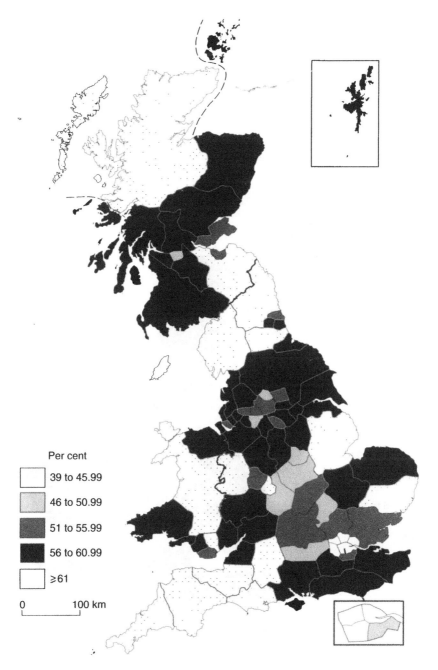

**Figure 3.6**  Percentage leavers into jobs, Job Centre Plus Districts, January 1999 to June 2003.
Source: New Deal indicators, Department for Work and Pensions.

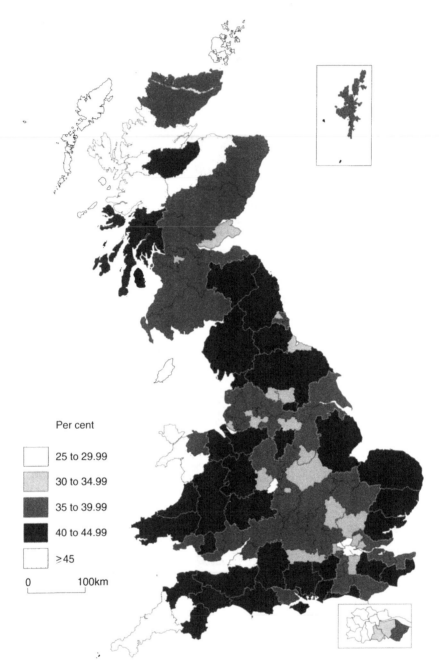

**Figure 3.5** Job-entry rate for all Units of Delivery to March 2002. Source: New Deal Indicators, Department of Work and Pensions.

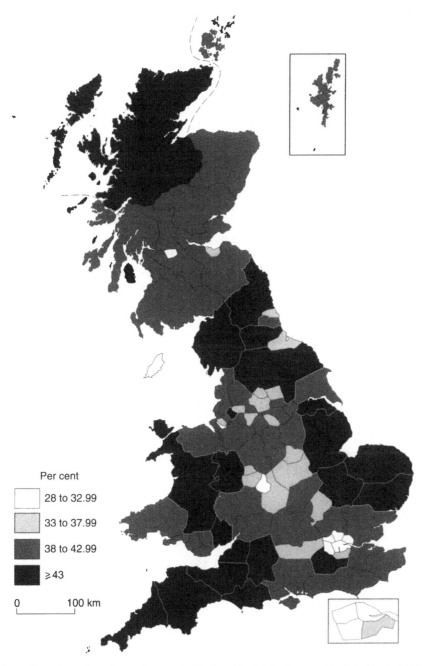

**Figure 3.7** Percentage leavers into unsubsidised sustained jobs, January 1999 to March 2003.
Source: New Deal indicators, Department for Work and Pensions.

Per cent

28 to 32.99

33 to 37.99

38 to 42.99

≥43

0          100 km

Employment Service classified UoDs into seven 'cluster types', on the basis of relative unemployment rates and population densities. These ranged from rural through rural/urban to urban, each of which was further subdivided into tight labour markets or high unemployment labour markets, with the addition of a further cluster of 'inner city high unemployment' (Cluster G) (see Martin et al., 2001). These proportions of clients finding subsidised and unsubsidised jobs clearly declined over the seven first cohorts (probably because later cohorts had a larger proportion of individuals who were re-entrants) (Table 3.1). But this decline did not involve a convergence between inner city high unemployment labour markets ('cluster G') and more buoyant labour markets. The somewhat unsurprising impression is that the New Deal has been least effective in inner city labour markets, that is in the areas where the youth unemployment was most acute. Further evidence in support of this claim, is apparent in the balance of the options or pathways chosen by New Deal participants. Recall that there were four options available (subsidised Employment, Full-time Education and Training, and the less popular Voluntary Sector and Environmental Taskforce placements). The take-up of the Employment option was

Table 3.1  Proportions of NDYP cohorts into 'all jobs' by UoD cluster type, at April 2000

| UoD Cluster Type | Cohort 1 | Cohort 2 | Cohort 3 | Cohort 4 | Cohort 5 | Cohort 6 | Cohort 7 |
|---|---|---|---|---|---|---|---|
| A (Rural tight labour markets) | 65.23 | 61.67 | 59.21 | 59.61 | 59.67 | 53.89 | 39.66 |
| B (Rural high unemployment) | 65.0 | 58.12 | 55.84 | 59.94 | 51.17 | 50.21 | 40.54 |
| C (Rural/urban tight labour markets) | 61.13 | 55.0 | 53.34 | 56.32 | 53.12 | 48.52 | 39.35 |
| D (Rural/urban high unemployment) | 60.4 | 57.04 | 54.71 | 58.87 | 55.39 | 49.55 | 38.58 |
| E (Urban tight labour markets) | 57.46 | 53.91 | 53.18 | 53.65 | 53.27 | 48.81 | 38.36 |
| F (Urban high unemployment) | 58.19 | 53.84 | 51.28 | 54.5 | 52.76 | 49.1 | 36.65 |
| G (Inner city high unemployment) | 47.04 | 44.94 | 41.59 | 43.14 | 41.71 | 36.91 | 28.05 |

Source: Employment Service Core Performance database.

lowest, and take-up of the Education and Training option highest, in the inner city, high-unemployment cluster of UoDs (Table 3.2).

As the Government itself has emphasised, the effectiveness of the programme should be judged not just in terms of people finding jobs but also in how well they retain them. Figure 3.8 shows the proportion of the first six cohorts of programme participants who remained in jobs 26 weeks after leaving New Deal at April 2000. The measure appears to correlate positively with the state of the local labour market: thus the highest rates of job-retention were to be found in the more buoyant and dynamic local labour markets in central and southern England. Conversely, job retention was lower in less buoyant peripheral and industrial local labour markets in northern England, Cornwall and Wales and Central Scotland. Again, most of the major conurbations had low rates, although the data suggest that the experience within London has been complex and diverse, including both areas with high job retention rates (such as Tower Hamlets, Sutton and Merton, Hounslow and Richmond) and areas with noticeably low retention rates (e.g. Lambeth and Bexley). There are likely to be several reasons for the better retention rates in regions in the south of the country. Areas of weak

**Table 3.2**  Balance of options by cluster type, first seven cohorts at April 2000

| UoD Cluster Type | Employment Option | Full-Time Education and Training Option | Voluntary Sector Option | Environmental Task Force Option |
|---|---|---|---|---|
| A (Rural tight labour market) | 24.46 | 37.14 | 20.24 | 18.16 |
| B (Rural high unemployment) | 25.16 | 35.44 | 17.19 | 22.21 |
| C (Rural/urban tight l. markets) | 18.08 | 41.68 | 21.51 | 18.72 |
| D (Rural/urban high unemployment) | 21.81 | 36.15 | 19.87 | 22.17 |
| E (Urban tight labour market) | 17.37 | 40.63 | 22.26 | 19.73 |
| F (Urban high unemployment) | 18.61 | 42.0 | 18.42 | 20.97 |
| G (Inner city high unemployment) | 14.09 | 49.57 | 21.05 | 10.52 |

Note: Figures show mean percentages of all options taken.

Source: Employment Service Core Performance Measure B.

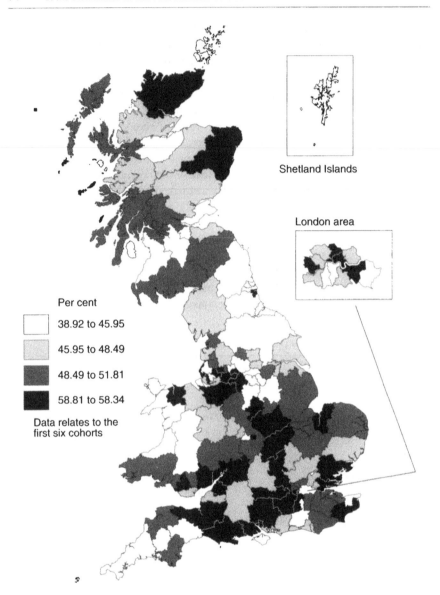

**Figure 3.8**  Proportion of participants remaining in jobs 26 weeks after leaving New Deal, at April 2000. Source: New Deal indicators, Department for Work and Pensions.

employment growth may well be marked by the presence of more insecure, 'hire and fire' jobs at the bottom of the skill structures in service and construction sectors. In addition, by boosting the supply of relatively less skilled labour in areas of low labour demand the programme may have served to exert downward pressure on wages for this sort of labour, which in its turn may increase labour turnover (Sargeant and Whiteley, 2000).

The proportion of the first seven cohorts of participants who left the New Deal for unknown destinations was also quite revealing (Figure 3.9).[4]

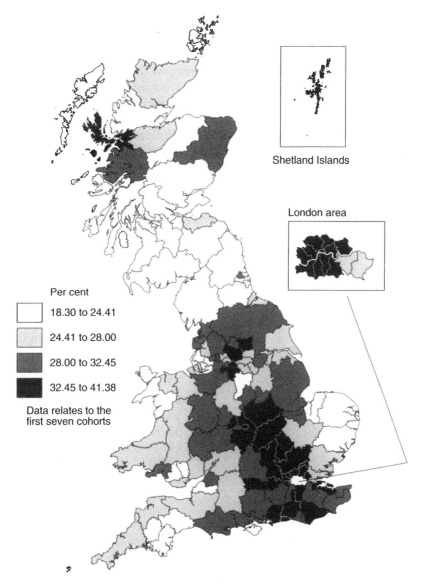

**Figure 3.9**  Proportion of participants leaving New Deal for unknown destinations, at April 2000. Source: New Deal indicators, Department for Work and Pensions.

These proportions were surprisingly high, although other research has found that about 60 per cent of these untraced movements in fact represent young people moving into jobs (Hales and Collins, 1999; DfEE, 2001b). This is certainly likely to the case in more buoyant labour markets. But, equally, these officially untraced flows may have different causes in different localities. It is noticeable that many of the large inner urban areas also have shown high rates of 'unknown destinations', and the reasons here may have as much to do with individuals disengaging from the policy and withdrawing from formal labour market participation (see Hoogvelt and France, 2000): as we noted in the previous chapter, youth economic inactivity rates vary considerably across the country. We also know that youth employment rates tend to be much lower in many inner city areas and more depressed labour markets, particularly in London, and it is one indicator of discrimination against ethnic minorities (see Table 3.3, Field-house et al., 2002a, 2002b). Several studies report that there has been little sign of a spatial convergence in (in)activity rates across the UK (Webster,

**Table 3.3**  Employment rates among 18–24 year olds for selected unitary authorities and counties, June–August 2000

| Unitary Authority/County | Employees and Self-Employed as Per cent of Base Population (18–24 Year Olds) |
|---|---|
| Hackney | 35.44 |
| Newham | 37.20 |
| Newcastle upon Tyne | 44.21 |
| Liverpool | 50.07 |
| Birmingham | 53.76 |
| North Tyneside | 61.18 |
| City of Glasgow | 61.70 |
| Dundee City | 62.56 |
| Manchester | 69.90 |
| Oxfordshire | 69.90 |
| Croydon | 71.32 |
| Cheshire | 73.58 |
| Leeds | 73.93 |
| City of Edinburgh | 75.89 |
| West Sussex | 77.50 |
| Bournemouth | 85.89 |

Note: Figures exclude young people on government training schemes.

Source: Labour Force Survey, Local Authority Database for 1997 and 1998.

2000). Furthermore, the decline in youth unemployment has not involved a levelling out or spatial convergence in its relative incidence across the country. Figure 3.10 shows youth unemployment rates by local authority district for 2003. Despite the impact of the NDYP, there continues to be an unmistakable 'north–south divide' in the relative incidence of youth unemployment (compare Figure 3.2, which shows the situation prior to the programme's start in 1997).

In summary, then, the overall weight of evidence from the Government's own performance measures, both for the periods up to and after 2000, suggests that the New Deal has not worked as well in the inner cities and the older industrial conurbations and cities as it has in other parts of the country. We now develop and substantiate this claim by examining some of the flows into and out of unemployment in a selection of UoDs during the first two years of the programme.

## Local Labour Market Flows

The local disparities in youth unemployment mapped above refer to *stocks*, that is to differences in relative proportions and rates. However, unemployment is not just about stocks, but also about *flows* into and out of joblessness: unemployment is not simply a state, but a *process*. In economics there has been an ongoing debate about the relative significance of unemployment inflows and outflows in determining the behaviour of the aggregate national unemployment rate. As we saw in the previous chapter, in the British case, some authors (and especially Layard – one of the protagonists of the New Deal) have argued that the rise and persistence of high unemployment during the 1980s was primarily due not to an increase in inflows into unemployment, but to a fall in the outflow rate and a consequential rise in the average duration of joblessness (see Layard et al., 1991). It is a fact that long-term unemployment, as a proportion both of the workforce and of the total unemployed, rose steeply in the 1980s and remained high for much of the 1990s. However, others, such as Burgess (1989), have contested this view, and have argued that the fall in the outflow rate and rise in duration were more *consequences* than causes of rising unemployment in the UK during the 1980s, and that rising inflows were in fact the main reason for the increase in unemployment in that decade. The issue is clearly of central importance, since the different views lead to rather different policy conclusions. For Layard (1997b), the need is for active labour market policies (ALMPs) that target the long-term unemployed and that increase their search effectiveness: the New Deal programme is of this sort. The Burgess-type view, on the other hand, suggests that the real challenge is to stem inflows into unemployment in

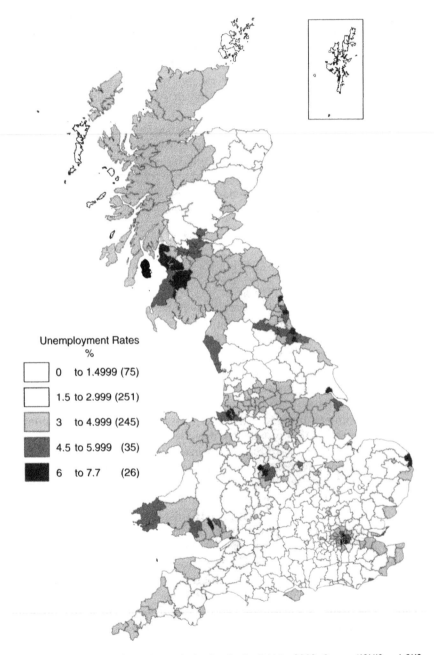

**Unemployment Rates**
%

| | |
|---|---|
| 0 to 1.4999 (75) | |
| 1.5 to 2.999 (251) | |
| 3 to 4.999 (245) | |
| 4.5 to 5.999 (35) | |
| 6 to 7.7 (26) | |

**Figure 3.10** Youth unemployment rates by local authority districts, 2003. Source: NOMIS and ONS.

the first place: that is, to operate as much on the demand side as on the supply side of the labour market equation.

Riley and Young (2000) argue that in assessing the implications of the New Deal for employment and unemployment it is vital to distinguish its effects on *flows* over a period of time from that on *stocks* at a point in time (see also Anderton et al., 1999). By comparing flows into and out of claimant unemployment for the 'client' age group (18–24 year olds) with that of other age groups they estimate the probable difference made by the NDYP. They argue that outflows from the register have increased substantially, especially for those unemployed for more than nine months. At the same time, they also note that the average inflow rate for the client group rose by 7 per cent in the two years after the New Deal was introduced, compared to the rate for 25–29 year olds.[5] This would amount to five and a half thousand additional inflows to unemployment per month. While this lends some support to the 'churning' idea, they conclude that 'The rise in the numbers flowing out of unemployment is almost twice as high as the numbers flowing into unemployment due to NDYP, so that the net impact of the programme is to reduce unemployment' (Riley and Young, 2000, p. 2). However, once again this macro-economic conclusion may mask significant local variations as local labour markets are frequently characterised by varying intensities of labour market transitions, constituting different 'flow regimes'.

As Martin and Sunley (1999) have shown, in one of the most detailed analyses to date, the different regions of the UK have experienced quite different unemployment 'flow regimes', that is different specific combinations of inflow and outflow patterns, and these different regimes are a major factor explaining regional unemployment rate disparities. More particularly, they find that during the 1980s and first half of the 1990s, the lower-unemployment, lower long-term jobless regions of southern Britain all had below average unemployment inflow rates and above average outflow rates. Conversely, most of the higher-unemployment, higher long-term jobless northern regions of the country had above average inflow rates and below average outflow rates. Moreover, regional differences in inflow rates – not outflow rates – have been the primary determinant of regional unemployment disparities. What this study suggests is that unemployment dynamics – in terms of relative inflow, outflow and duration – can vary considerably from local area to local area. This in turn implies that, perhaps more emphatically than has been officially recognised, the specific task confronting the New Deal has differed in different parts of the country, with the implication that its impact is likely to have varied spatially as a consequence.

To examine these issues, we turn our attention in this section to an analysis of five case-study UoDs – Cambridge, Camden and North

Islington, Birmingham, Tyneside North, and Edinburgh and East and Mid Lothian. While these have been selected on a combination of theoretical and pragmatic grounds, they provide examples of different types of labour market in different regions of the country, as well as different New Deal delivery models (some of the salient features of these case study UoDs are given in Chapter 4: these same local labour markets were also used as the locations for intensive surveys and interviews with New Deal managers, participating employers and young unemployed – see Nativel, Sunley and Martin, 2003). We examine some of the flows into and out of claimant unemployment in these UoDs, together with 13 other UoDs characterised by strikingly different labour market conditions. Unfortunately we have not been able to include rural UoDs in this analysis of inflows and outflows as it is impossible to calculate reliable inflow rates for these areas.[6] In general, the aim is to compare some of the most prosperous and dynamic local labour markets in Britain – such as Cambridge, Crawley and Oxfordshire – with some of the most problematic and distressed, typified by Liverpool, Glasgow, Birmingham, and Hackney (London). In addition, we have included some urban areas that have experienced substantial employment growth – Edinburgh, Leeds and Peterborough – in order to be able to comment on the full range of urban experiences. Claimant unemployment flows, of course, yield only a partial picture of the conditions in these local labour markets, as they provide no clues as to what is happening to participation rates or those wanting work.[7] Nevertheless, for our purpose here, the claimant flows are useful in providing an insight into the different problems faced by local New Deal Agencies as well as some of the possible local effects of the programme.

Figures 3.11 and 3.12 show the stock of the long-term unemployed aged 18–24 for our selection of 18 UoDs. These figures show quarterly averages from the first quarter of 1997 to the second quarter of 2000. What is clear is that the stocks were falling in all of the selected labour markets prior to the introduction of the New Deal programme, so that it was clearly coming into operation in the context of a strongly favourable cyclical trend in the national economy. Although this cyclical trend became more attenuated during 1999, there is no question that a substantial decline in long-term youth unemployment has occurred in all of these very different types of local labour market. At the national scale, Riley and Young (2000) have estimated, using econometric models, that without the New Deal long-term youth unemployment would be about double current levels. But this is unsurprising given that there is now no option of continuing on unemployment benefit indefinitely after the initial Gateway stage of four months. Thus the main reason for these declines in stocks has been an increase in outflows from the claimant register. Figures 3.13 and 3.14 show the quarterly outflow rates, as a percentage of the total claimant stock, for

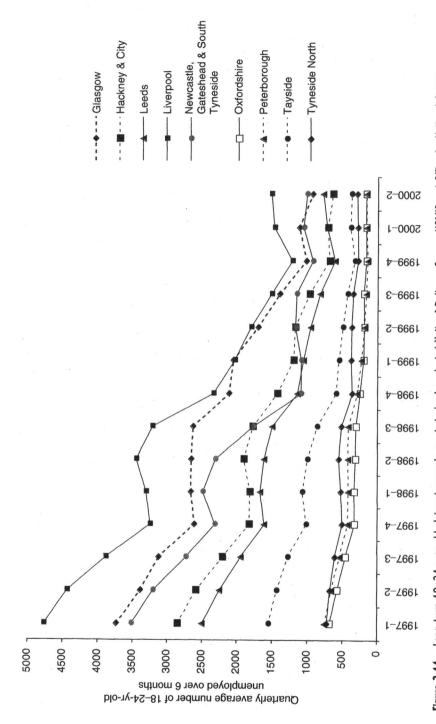

**Figure 3.11** Long-term 18–24 year old claimant unemployment stocks, by selected Units of Delivery. Source: NOMIS and Office for National Statistics.

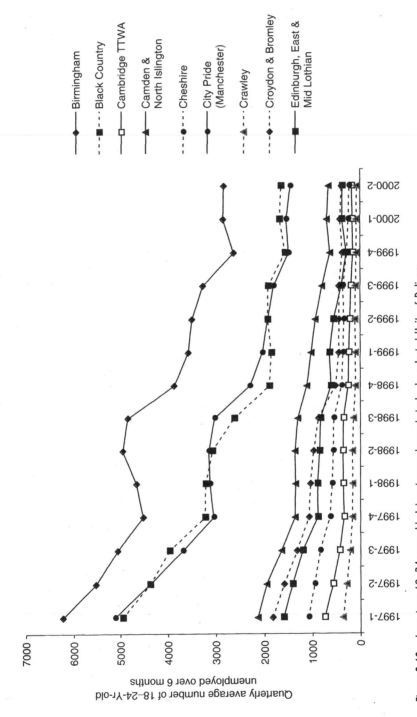

**Figure 3.12** Long-term 18–24 year old claimant unemployment stocks, by selected Units of Delivery.

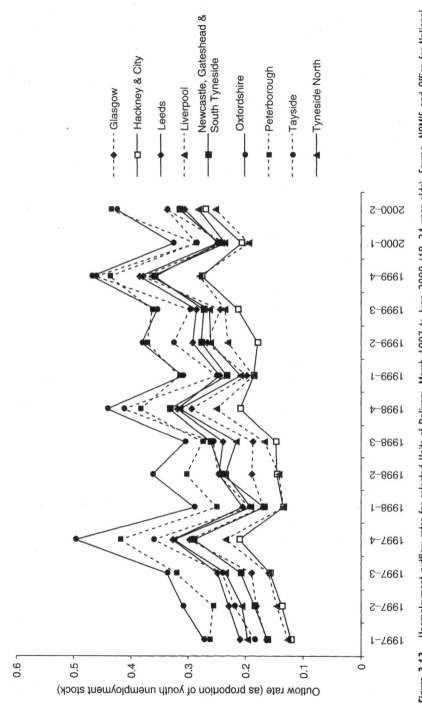

**Figure 3.13** Unemployment outflow rates for selected Units of Delivery, March 1997 to June 2000 (18–24 year olds). Source: NOMIS and Office for National Statistics.

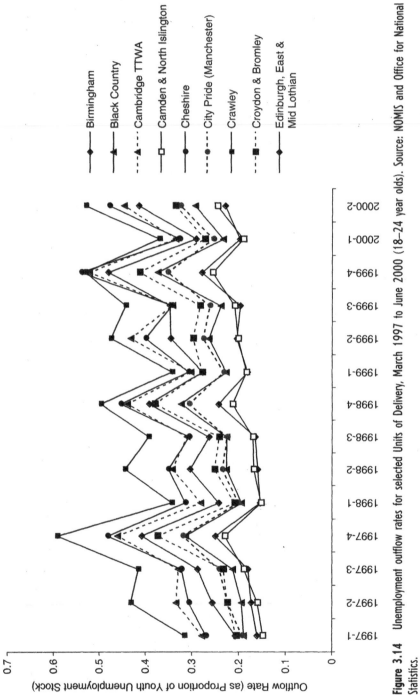

**Figure 3.14** Unemployment outflow rates for selected Units of Delivery, March 1997 to June 2000 (18–24 year olds). Source: NOMIS and Office for National Statistics.

each of our selected UoDs. As would be expected, there are large seasonal variations in the outflow rates, but underlying these variations is a perceptible trend for outflow rates to rise. But what is also evident is that there are significant and systematic differences in outflow rates across different UoDs, with those in tight labour markets such as Cambridge, Crawley, Cheshire, Peterborough and Oxford being up to double those found in the relatively depressed areas such as Glasgow, Liverpool, Birmingham, Newcastle and Hackney. In general, there have been few shifts in the hierarchy evident at the start of the period. The rising trend is clearer in Table 3.4 which shows the average rates for the New Deal UoD 'cluster groups'. While all groups have benefited from increasing outflow rates, the group of inner city high unemployment UoDs (cluster 'G') has clearly not improved relative to the other types.

Other research suggests that there may be differences between areas in the composition of the destinations of outflows from unemployment. Even at a broad regional scale there is evidence that the programme's effects on outflows have varied with the level of demand for labour. McVicar and Podivinsky (2003) used a 5 per cent sample of all JSA claimants between 1996 to 2001 to examine the NDYP's effects on probabilities of exit from claimant unemployment across the regions. They found that in most regions the primary effect was to shift young people unemployed for more than six months into education and training rather than into jobs, and that the relative size of this effect tended to be positively correlated with initial unemployment rates. They conclude, 'NDYP shifts more young people into employment and less young people into education and training in low unemployment areas than in high unemployment areas' (p. 28).

However, outflow rates are obviously only half the story. We have also attempted to calculate inflow rates for the same UoDs. The difficulty here is that of obtaining an estimate of the source population (the denominator)

Table 3.4   Outflow rates by cluster, quarterly averages for second quarter in each year, 1997–2000

| Cluster Type | 1997 | 1998 | 1999 | 2000 |
|---|---|---|---|---|
| A (Rural tight labour markets) | 0.315 | 0.337 | 0.399 | 0.463 |
| B (Rural high unemployment) | 0.291 | 0.343 | 0.415 | 0.487 |
| C (Rural/urban tight labour markets) | 0.305 | 0.341 | 0.397 | 0.439 |
| D (Rural/urban high unemployment) | 0.246 | 0.250 | 0.323 | 0.368 |
| E (Urban tight labour markets) | 0.274 | 0.289 | 0.349 | 0.395 |
| F (Urban high unemployment) | 0.223 | 0.240 | 0.314 | 0.335 |
| G (Inner city high unemployment) | 0.169 | 0.183 | 0.241 | 0.273 |
| GB Average | 0.283 | 0.308 | 0.377 | 0.429 |

for calculating the inflow rates on to the register, that is young people starting to claim benefit. The only statistics available for this particular period are the total populations of 18–24 year olds for 1997 and 1998 for local authority districts (LADs).[8] In order to estimate a source population, we have taken an average of the two years and amalgamated and allocated LADs into respective UoDs. As the population figures are derived from sampling, in several rural LADs the population totals are too small to be reliable and we have therefore been unable to calculate reliable inflow rates for all UoDs, or averages for the different New Deal 'cluster groups'. However, on this basis it has been possible to estimate inflows for our selection of local labour markets.[9]

Figure 3.15 reveals a predictable hierarchy with the Manchester UoD showing a stable inflow rate of about 3 per cent, Birmingham and the Black Country at 2 per cent and with Crawley, Cambridge and Croydon and Bromley stable or declining slightly at around 1 per cent. Thus the local labour markets with the highest outflow rates tend also to be those with the lowest inflow rates. The contrasts in Figure 3.16 are even more striking and show that the overall dispersion in inflow rates was actually increasing. The UoDs with the highest inflow rates, Hackney, Newcastle and Liverpool, appear to have had slight increases over the period. Most of the UoDs showed stable trends, but in the most buoyant labour markets (Crawley and Oxfordshire), the inflow rates appeared to decline. As might be expected, their tight labour markets meant that fewer young people were entering the unemployment register. However, in most of the higher un-employment urban areas, the inflow rates were stable and in a minority of urban areas they were actually increasing. This is quite surprising in the context of the general climate of strong employment growth over the late 1990s (in the UK, the general behaviour of the labour market is such that on-flows to unemployment tend to rise in recessions and fall in booms). One possible explanation is that an expansion of employment encourages some young people to quit their jobs and claim benefit while they look for a new ones. But this is unlikely, as this effect would be expected to be strongest in buoyant labour markets such as Crawley and Oxfordshire. Another possible explanation is that those labour markets with rising in-flows also have rising employment rates as young people are drawn back into labour force participation, but this does not explain by itself why inflows to unemployment should also rise, unless of course these jobs are proving to be short term and temporary.

It is much more likely that the variations in inflow rates were a reflection of the deficiencies in the demand for labour in many industrialised urban and inner city labour markets. As several authors have argued, when the surplus labour force is defined to include all those wanting work,[10] there is evidence of a severe underlying 'jobs gap' in these areas (Turok and Edge,

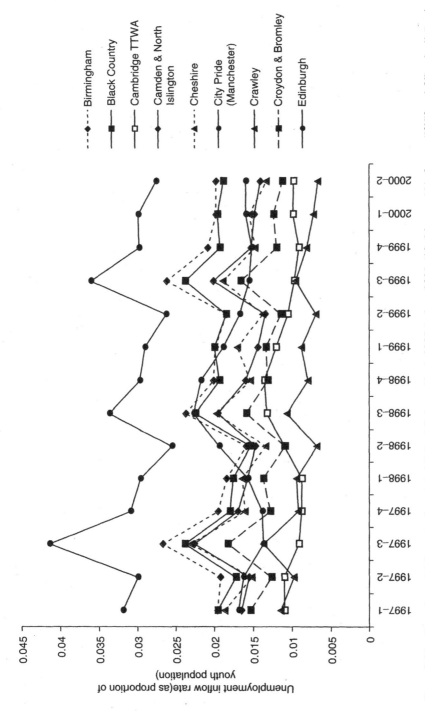

**Figure 3.15** Unemployment inflow rates for selected Units of Delivery, March 1997 to June 2000 (18–24 year olds). Source: NOMIS and Office for National Statistics.

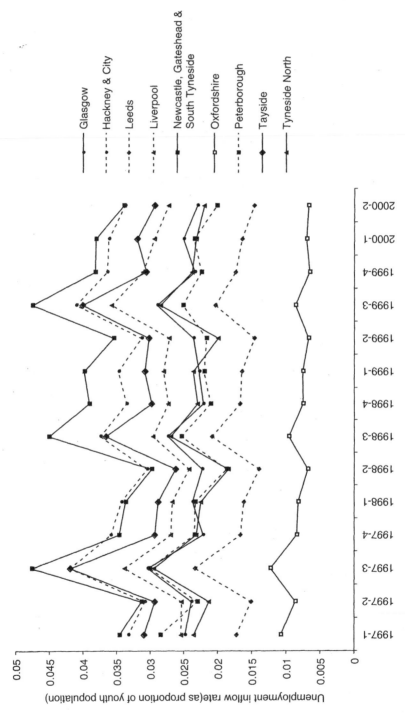

**Figure 3.16** Unemployment inflow rates for selected Units of Delivery, March 1997 to June 2000 (18–24 year olds). Source: NOMIS and Office for National Statistics.

1999; Webster, 2000; Glyn and Erdem, 1999; Evans et al., 1999), and this would, of course, tend to raise unemployment inflow rates. Furthermore, in areas such as Hackney, the problem is compounded by the added difficulties of some ethnic minority groups in attaining any available jobs. It is probable that the trends in inflow rates in slack labour markets indicated a certain recycling of individuals from the New Deal programme back into unemployment. In the context of the economic climate prevailing in this period, the scale of change in numbers of inflows is very difficult to explain without accepting the existence of a 'revolving door' or 'churning' between New Deal options and benefit claiming. In Chapter 4, we examine this issue in more detail.

## Local Disparities in Other New Deals

Thus far, then, we have argued that our understanding of the degree of success of the NDYP needs to pay careful and close attention to the geography of its outcomes, and that it is too simplistic to summarise and downplay its geography by invoking the persistence of residual pockets of local worklessness. It is also important to consider whether the degree of geographical variability shown by the NDYP is exceptional or whether marked geographical variations have been visible in the outcomes of other New Deal programmes. To what degree is there something inherent in New Deal approaches that is unable to cope with the varying conditions across Britain's local labour markets? Here we select the New Deal 25 Plus and the New Deal for Lone Parents as there is some information available from official reports and databases on their regional and local outcomes.

There is a consensus that, in terms of key performance indicators, the New Deal for the long-term unemployed aged over 25 has been much less successful than the NDYP. The original New Deal for the long-term unemployed introduced in June 1998 had much lower levels of funding per client than the NDYP (funding per client was six times greater on the latter), it lacked a Gateway and instead offered only an advisory interview process lasting between three and six months and involving a series of mandatory interviews with New Deal personal advisers (Lindsay, 2002). Participation in other activities was voluntary and in most areas of the country (excepting a set of pilot areas) individuals had to be unemployed for two years before they became eligible. It is not surprising, then, that by the start of 2001 only 15 per cent of 330,000 New Deal 25 Plus clients had found employment and only 12 per cent had obtained sustained jobs (Lindsay, 2002). Evaluations reported that nearly all of these would have found work in the absence of the programme (Wilkinson, 2003). In April 2000 the programme was enhanced to introduce a short Gateway period, and in April 2001 the programme was re-engineered to incorporate many of the features found in the NDYP. Those unemployed for 18 months would now be eligible for entering the programme. It was given a full Gateway and those unable to find work during this period would undertake a compulsory Intensive Activity Period (IAP) of between 13 and 26 weeks. This IAP incorporates many of the NDYP option activities including a subsidised work placement, work experience, basic employability training, self-employment support and work-focused training (Wilkinson, 2003). In this way the programme was reformed in order to replicate the individual and intensive style of support of the youth programme, in the hope that it would help individuals to target opportunities in growing sectors of the economy (Lindsay, 2002).

The reform and enhancement of the ND 25 Plus appears to have improved its aggregate results. Of the 290,950 people who started since April 2001, 23.7 per cent had moved into sustained jobs by the end of September 2003, and 30.2 per cent had moved into a job of some kind (Bivand, 2003).[11] There appears to have been improvement in the results over time, although the labour market context has itself been improving. On the original and enhanced programmes over half of all leavers were returning to JSA, while for the re-engineered programme, this figure fell to about half that, although the number leaving to other benefits has increased slightly, possibly because the IAP is now mandatory and a significant number of clients drop out at the start of the IAP (see Chapter 4).[12] Official evaluations of the ND 25 Plus have examined differences in outcomes for different groups in some detail and have established that the percentage of leavers going into unsubsidised employment declines with age. However, there has been little examination of the geographical disparities in outcomes.

This neglect is remarkable given the striking differences in performance between Districts (see National Employment Panel, 2004). Figure 3.17 shows that there has been an inverse relationship across areas between leaving to take jobs and leaving to return to JSA claiming. The differences on both indicators are very marked. The lowest rates of leaving into jobs are in London Central South (17.6 per cent), Renfrewshire (19.7 per cent), London East (20.6 per cent), Hull and East Riding (20.8 per cent) (See Table 3.5). In some contrast the highest job destination rates are in Grampian (38.5 per cent), Wiltshire (34.4 per cent), East Lancashire (33.2 per cent). Some of the low job attainment districts also have very high rates of returning to JSA (including Eastern Valleys (39.1 per cent), London Central South (38.0 per cent), London East (36.4 per cent), Hull and East Riding (34 per cent), Renfrewshire (33.3 per cent) and Glasgow (28 per cent). Again, in contrast, the lowest rates of leavers returning to JSA are in Dorset, Wolverhampton and Walsall, and Wiltshire with rates just above

**Table 3.5** Immediate destinations for New Deal 25 Plus to December 2003

| *Worst Performing Districts* | | | *Best Performing Districts* | | |
| --- | --- | --- | --- | --- | --- |
| *JCP District* | *% Leavers to Jobs* | *% Leavers Returning JSA* | *JCP District* | *% Leavers to Jobs* | *% Leavers Returning JSA* |
| London Central South | 17.6 | 38.0 | Grampian Moray | 38.5 | 23.0 |
| Renfrewshire, Inverclyde | 19.7 | 33.3 | Wiltshire | 34.4 | 22.7 |
| London East | 20.6 | 36.4 | East Lancashire | 33.1 | 24.9 |
| Hull and East Riding | 20.8 | 34 | South Humberside | 32.8 | 28.0 |
| London South-East | 21.0 | 37 | Dorset | 32.8 | 20.2 |
| Lanarkshire | 21 | 29.0 | Cambridgeshire | 32.0 | 30.4 |
| London Central North-West | 21.1 | 33.6 | Devon | 31.5 | 25.3 |
| Sheffield | 21.2 | 32.3 | North-West Wales Powys | 30.9 | 25.5 |
| Fife | 21.6 | 31.6 | West of England | 30.7 | 24.7 |
| Eastern Valleys | 21.7 | 39.1 | Edinburgh | 30.4 | 25.9 |
| Wrexham North Wales | 21.8 | 30.7 | Hampshire | 30.3 | 28.1 |
| West Wales | 22.4 | 31.2 | Cheshire Warrington | 29.5 | 23.3 |
| Ayrshire Dumfries Galloway | 22.4 | 32.3 | Hereford and Worcester | 28.6 | 25.2 |

Source: New Deal statistics, Department for Work and Pensions.

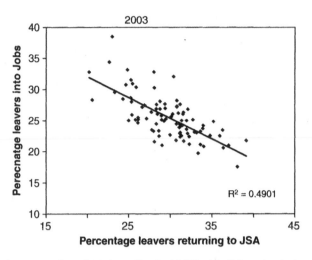

**Figure 3.17** The inverse relationship between New Deal 25 Plus job attainment and returns to JSA by JCP District, January 2002 to December 2003. Source: New Deal indicators, Department for Work and Pensions.

20 per cent. To summarise, with some exceptions, there is a clear pattern to these performance outcomes with rural Districts and prosperous labour markets generally performing up to twice as well as many traditional industrial urban areas and Districts in inner London.

The broad pattern is confirmed if we examine the proportion of Enhanced ND 25 Plus participants obtaining unsubsidised and sustained jobs over the two-year period ending in December 2003. Again, the lowest rates are in Eastern Valleys (16 per cent), London Central (16.6 per cent), Glasgow (17.6 per cent) Doncaster (18.6 per cent), Renfrewshire (19 per cent), London East (19.2 per cent), Hull and East Riding (19.3 per cent), Sheffield (19.4 per cent) and Lanarkshire (19.8 per cent). The highest rates are again up to double these. On this indicator the programme is working most effectively in Grampian (35.9 per cent), Wiltshire (34.6 per cent), Dorset (31.6 per cent), South Humberside (31.3 per cent), East Lancashire (31.2 per cent), Cambridgeshire (30 per cent), North-West Wales (30 per cent), Devon (29.7 per cent) and Edinburgh (29.4 per cent). While there may well be variations in managerial practices between JCP Districts, as well as possible variations in statistical efficiencies, affecting these results, the consistency of findings on the different indicators strongly indicates that the programme is being significantly influenced by the character and extent of problems in the local labour supply and by the level of current demand in the local labour market. Given these results, it would be difficult to argue that contextual factors do not have a profound impact on the degree of programme effectiveness.

Further, there is evidence that the outcomes of the New Deal for Lone Parents have also been markedly shaped by local labour market contexts. This programme differs from the youth and 25 Plus programme of course in that only the initial work-focused interview is mandatory and further participation is voluntary. However, a recent review of the New Deal for Lone Parents identified strong locational factors and large regional differences in outcomes with only 37.5 per cent of leavers from the programme in London entering employment between October 1998 and November 2002, compared to 58 per cent in Wales and 56 per cent in Yorkshire and Humberside (Evans et al., 2003, p. 79). If job attainments are expressed as a percentage of initial interviews over the period October 1998 to September 2003, the figures ranged from 29 per cent in London to 44 per cent in Wales and Yorkshire Humberside (Bivand, 2004). While these figures suggest that the programme in London has encountered specific problems and obstacles, variations at district level are probably more revealing.

Variation in employment outcomes at district level is even greater and is increasing over time. Participants in rural areas also seem to have lower probability of leaving the programme and entering work. The highest rates of job entry are commonly between 74 to 80 per cent while lowest level outcomes are around 30 per cent. Preliminary analysis by the DWP has apparently found that there are strong associations between district performance and female wage rates and population density, but that environmental factors accounted for only a quarter of the total variation (Evans et al., 2003). The methodological basis for this conclusion is not explained, however. The report also admits that the probability of gaining employment from the programme is also associated with ward-level deprivation and that participants in the least deprived 25 per cent of wards have higher probabilities of gaining work. In a telling phrase Evans et al. write:

> Some lone parents live in areas where suitable jobs and childcare are readily available and accessible, others live where neither are easy to find. NDLP by itself cannot do much to improve the number of jobs or childcare places, nor to improve local transport services but it can provide better local knowledge and accurate information to lone parents, so that they know exactly what is available in their local area (pp. 105–106).

One of the main constraints on the programme has been the restricted range and low quality of many if the jobs obtained by participants. The relative lack of skills and experience of many lone parents means that in depressed labour markets participants are tending to gain only low paid and precarious, insecure jobs (Casebourne, 2003). The insecurity and low rewards of these jobs is contributing to significant recycling between work and the programme.

The programme's eventual results are clearly dependent on the capacity of local labour markets to absorb a large number of new workers into mainly entry level jobs (Millar, 2000b), and this has important implications for debates on the future direction of the programme. The level of compulsion on the NDLP has been increased with the introduction of mandatory work focused interviews and there has been a lively debate on whether further increasing the level of compulsion is desirable. This is not only because pushing the work-first agenda for lone parents undermines the value of unpaid caring responsibilities (Sumaza, 2001), but also because the implications will be very different depending on local context. As Millar argues:

> Much depends on the nature of the compulsion, the context in which it is operated, and the type of sanctions used. Being required, under threat of loss of benefits, to take any job regardless of individual circumstances, of the pay and condition of that job, and of whether adequate childcare is available is rather different from being required to seek employment and/or training in a buoyant and well-paid labour market, where childcare is readily available and widely used (2000b, p. 341).

## Conclusions

We began this chapter with the official positive story that has been told about the effectiveness of the New Deal. In general this has downplayed geographical variations and argued that the programme's success and the retreat of unemployment have exposed localised concentrations of the most unemployable, concentrated in residual pockets where the local labour supply is very poor and multiply disadvantaged. There are clearly some elements of truth to this evaluation. Many deprived urban labour markets in the UK undoubtedly suffer from a severe lack of basic employability and shortage of basic skills, while in cyclical terms the youth labour market has been in a relatively healthy condition in recent years. However, the analysis in this chapter has shown that the problems of the New Deal programme are not adequately represented as being confined to localised pockets. Rather there has been a much wider problem of a lower effectiveness in urban areas relative to rural areas, and that this divergence has been especially marked in the largest cities and cities in old industrial regions. It is impossible to escape the conclusion that the path dependence of depressed labour market conditions in the country's urban and older in-dustrial areas has been manifested as lower job attainment rates, lower job retention rates, and generally static and, even in some cases rising, inflows to youth unemployment. The overwhelming weight of evidence leads un-avoidably to the conclusion that the impact of the New Deal for Young

People has varied significantly as between different local labour market areas across Britain, and in particular has been noticeably less effective in many inner urban and depressed industrial labour markets. In other words, the geography of programme outcomes has shown significant levels of variability, reflecting structural features of British labour markets. Moreover, the problem of geographical disparities in the achievement of aims is by no means confined to the NDYP. The New Deal for 25 Plus is now showing just as much, if not more variability in its performance across districts. While there are some locational factors specific to individual programmes, such as the importance of the availability of childcare to the NDLP, there is a broad shared pattern to the geography of results. To put it simply, the New Deal workfare programmes are far less successful in parts of London, in distressed labour markets and traditional industrial areas.

This problem of local differences in effectiveness is a worrying issue, given that one of the central elements of the new workfarism is its claim to respond much 'more effectively and flexibly' to different local labour market circumstances (see OECD, 2001). One problem is the way in which these differences in local labour market conditions are viewed and conceptualised by government policy-makers (and those economists who advise them). The distinctive feature of the new generation of workfare-type programmes, and other active labour market policies, is that they focus on the supply-side of the labour equation, and leave the demand-side 'to the market'. The underlying premise is a twofold one. First, that governments are unable to create jobs: all they can do it promote the 'macro-economic conditions' favourable for job creation (such as low inflation, low taxes, a stable currency, competition, enterprise, etc.). Second, where governments can intervene is by promoting improvements in the supply of labour – that is by increasing its 'employability'. The assumption is that an improved labour supply will create its own demand. Local differences in unemployment and joblessness are thus interpreted primarily as arising from supply-side failures and shortcomings, not from a local lack of demand for labour. Certainly, there is no doubt that supply-side measures like the New Deal are needed. But the question of local labour demand – the volume and range of jobs available – is also vital, and cannot be assumed to be non-problematic or self-correcting. Supply does not necessarily create its own demand: and local labour markets do not simply 'clear' even if supply-side weaknesses are removed (see Martin, 2000). So while we agree that depressed labour markets have severe and chronic problems of labour supply we would argue that we can not properly understand, nor address, these problems by focusing on supply alone. The next chapter examines some of the interactions between supply and demand in more detail by looking at local employability and the different experiences of the NDYP in different types of local labour market.

# Appendix

This appendix defines some of the institutional terms used in the chapter.

## New Deal Units of Delivery

The New Deal was initially administered through 144 local partnerships or programme areas. The geographical boundaries of these units were intended partly to encourage the formation of local programme delivery partnerships and partly to reflect travel-to-work patterns. The boundaries of the local 'units of delivery' (UoDs) of welfare-to-work are obviously important. As Hughes (1996) argues, if local programme area boundaries are drawn too tightly they can underbound the opportunities available to participants in nearby areas with higher concentrations of jobs. On the other hand, if they are drawn too widely, programme areas can obscure the existence of marked localised variations in unemployment and job availability. Travel-to-work distances have a particularly significant impact on the labour market opportunities for the unemployed and for low-skilled, low-paid workers, a fact that is recognised by the provision of certain travel to work allowances within the New Deal programme. An appropriate administrative geography, preferably one that bears a close approximation to actual local labour markets, is thus essential to solve the employment needs of welfare-to work. The New Deal local UoD areas appear to have been largely based on pre-existing Employment Service districts (aggregations of numerous local wards), although some districts have been amalgamated. Further, they vary considerably in geographical size: for example, from individual local boroughs in London to whole counties, as in Cheshire, Cumbria, or even sub-regions, as in Dumfries and Galloway. While this variation reflects the general geographical distribution and concentration of the working population and employment across the country, the UoDs did not necessarily correspond to meaningful functional labour market areas, and problems of underbounding and overbounding undoubtedly existed.

## Job Centre Plus Districts

From 2001 as part of the reorganisation of the Employment Service and its merger with the Benefits Agency, New Deal Units of Delivery were superseded by 90 Job Centre Plus Districts. These are composed of former Job Centre Areas and are larger than UoDs. The DWP has not been able to provide an official map of these Districts and we have therefore drawn up our own estimated and provisional maps (see Figures 3.6 and 3.7).

## New Deal for Young People core performance measures and key indicators

The Employment Service initially produced a wide-ranging series of nine core performance measures for each Unit of Delivery. These included (A) the numbers of participants and proportion of each cohort moving into (1) unsubsidised jobs, (2) unsubsidised jobs and, (3) all jobs; (B) the numbers of participants and the proportion of each monthly cohort moving from the Gateway and each of the Options into unsubsidised jobs; (C) the unit costs of these outcomes; (D) the number of participants and the proportion of each monthly cohort remaining in jobs thirteen weeks, six, twelve or eighteen months after leaving New Deal, as measured by the renewal or otherwise of claims for Job Seeker's Allowance or other benefits; (E) the numbers and proportions of participants who are disabled, from ethnic minority backgrounds and who are mean and women achieving the outcomes above; (F) the numbers of subsidised jobs made available by employers and the level of employer satisfaction; (G) the level of satisfaction among participating young people; (H) the number and level of qualifications achieved by participants; (I) the number and level of participants and the proportion of each monthly cohort leaving the New Deal for known destinations. In 2002 as part of the introduction of Job Centre Plus these core performance statistics were replaced by a series of NDYP Key Indicators for JCP Districts that show monthly and annual figures rather than outcomes for quarterly entry cohorts. These are Key Indicator 1, Job entrants as a proportion of all leavers; Key Indicator 2, Proportion of New Deal leavers who left for Sustained Unsubsidised Jobs; Key Indicator 3, Parity of Outcomes between White and Ethnic Minority Groups; Supplement 1, Job Entrants as a Proportion of All Leavers; Supplement 2, Proportion of New Deal leavers by stage and by length of time on New Deal who left into sustained unsubsidised jobs.

## Local programme delivery/management models

The different types of local partnership that have been established for the implementation of the New Deal all depend on an underlying contractualism in which money is given to service providers on the basis of the number of clients referred. This blend of contractualism with partnership-building allows a wide range of agencies to be involved and used (see Chapter 6). Four types of delivery models, based on different forms of contracting have been devised. In the most common, the Employment Service led model, the Employment Service issues a series of contracts to individual New Deal providers for some aspects of the Gateway and each of the four workfare options. In some cases there may be a lead provider for a part of the programme and they may then subcontract to other agencies. The Consortium-led model, on the other hand, brings together a set of partners with a single lead organisation. In this case, the Consortium is accountable for the delivery and the ES is not part of the partnership: it simply contracts delivery to the Consortium as a whole, which is nominally responsible for choosing and monitoring the providers. The Joint Venture Partnership model is similar to the Consortium model but here the ES is an equal partner. Finally, the fourth model (which only applies to ten UoDs) is the private sector-led model, in which a private sector organisation is contracted to be responsible for the delivery of the whole programme. When the New Deal was introduced, one of the delivery models was set up in each of the 144 UoDs across the country.

# Chapter Four

# Welfare-to-Work in Local Context

## Introduction: Employability and Local Context

The rise of active labour policy has given an associated prominence to the concept of 'employability', and one of the key aims of the New Deal has been to raise the employability of its participants. While the concept is often used loosely and with different meanings, most definitions agree that it refers to the capability to both obtain and maintain sustainable and fulfilling jobs (Hillage and Pollard, 1998; McQuaid and Lindsay, 2002). It signals the need for individuals to develop and deploy transferable skills in order to operate effectively and to progress in the context of an increasingly insecure and flexible job market. In a frequently quoted report, Hillage and Pollard (1998) argued that 'employability' could be seen as being produced by the interaction of different factors. These include 'baseline employability assets', such as basic skills and personal attributes, and job specific generic and key skills; 'presentation skills', focusing on the demonstration of assets and strategic pursuit of opportunities; 'deployment capabilities', including job search and career management skills; and finally 'context factors', understood as the interaction of personal factors and the nature of the labour market. In this view, the ability to realise assets and skills will depend on the interaction between contextual socio-economic factors, such as local labour market demand and employer perceptions, and personal circumstances, especially caring responsibilities and individual health. Employability is a function of the degree of match between the demand for, and supply of, labour, in both quantitative and qualitative terms (Kleinman and West, 1998).

An inductive, empirically based typology to understanding the factors shaping employability, focused more specifically on the risks of non-employment, has been developed by Berthoud (2003). In a large-scale quantitative study of the risk of labour market exclusion based on Labour

Force Survey data, Berthoud identifies six sets of factors that significantly increase the risk of non-employment. Risk of non-employment is greatly increased for those aged 50 years or more; individuals without a partner and for people with children; those with poor qualifications; those with any health impairment; those belonging to a minority ethnic group; and individuals resident in areas with a lower demand for labour. The effects of each sets of variable is additive so that their combination significantly increases the risk of non-employment.

While these static compositional studies devote most of their attention to supply-side determinants of employability, they are clearly helpful in providing a framework for starting to understand both employability and non-employability. They suggest that labour market policies are correct to target certain supply-side features, but that such policies have major limitations if they overlook the interaction of supply with labour market demand and contextual factors. In terms of understanding the long-term evolution and dynamics of employability, this interaction is especially important, and policies that focus only on improving individuals' employability are unlikely to succeed. As we discussed in Chapter 2, the contemporary predeliction for supply-side interventions in the labour market either assume (or assert) that there is no shortage of labour demand ('there are plenty of unfilled vacancies'), or put undue faith in a sort of Say's law of the labour market, whereby it is assumed that improving labour supply will somehow of itself increase the demand for labour. In a dynamic longer-term view, an individual's employment experience and history are not separate from the local labour market(s) in which they are resident: the formation of employability is to a large degree relational so that an individual's employability is not independent of the local labour market context in which he or she seeks work, and cannot be properly understood in isolation from this context (Evans et al., 1999). Local context, education, socialisation and parental experience of work, shape the rate of accumulation and development of an individual's employability assets and capabilities. What this means is that local labour markets typically have different likelihoods and forms of non-employability and that the key forms of non-employability across different localities will vary, in part because of geographical differences in labour demand conditions. As McQuaid and Lindsay argue (2002):

> In buoyant labour markets . . . the personal characteristics of many individuals, and their awareness of job opportunities and the priorities of employers, will decisively impact upon their ability to find work. In those areas where demand deficiency is a problem a lack of opportunity will continue to limit the employability of many residents, whatever measures are taken to improve individuals' employability assets and their presentation and deployment (p. 626; see also Lindsay et al., 2003).

The precise make-up of the causes for non-employment in higher-demand local labour markets is likely to be different from that in low-demand labour markets.

To explore the ways in which the problems of low employability and obstacles to labour market participation faced by the NDYP vary between different local labour markets, this chapter uses material from case studies of five contrasting local labour markets. While the programme faced challenges even in dynamic and buoyant labour markets, in more depressed urban labour markets there have been severe limitations of a different order and kind. Our case studies reveal some of the ways in which a lower local demand for labour conditions the operation and impact of a workfare programme like the NDYP. The chapter argues that the problems faced in depressed labour markets have been distinct from those in areas of higher employment demand and that this has given rise to a worrying set of issues. One is the prevalence of forms of welfare recycling in which clients go through the programme and then quickly return to it again. Another is the diversion of New Deal participants into inactivity and benefits other than the Job Seekers' Allowance. The chapter starts by reviewing some of our case study findings in contrasting types of local labour market. It then uses more extensive statistical material to document some of the New Deal's limitations and unintended outcomes in the most problematic and depressed local economic situations.

## Low Employability in Buoyant Labour Markets

In order to investigate the problems faced in different local areas and the degree to which the NDYP was able to respond, we selected five contrasting local areas: Camden and North Islington, Birmingham, Cambridge, Edinburgh and North Tyneside. The five UoDs were selected to represent a range of different local labour market conditions, as well as the different New Deal delivery models outlined in the previous chapter (see Table 4.1). Three high-unemployment urban labour markets (Tyneside, Birmingham and Camden) were contrasted with two buoyant labour markets (Cambridge and Edinburgh). In each location, detailed interviews were carried out with managers, New Deal personal advisers (NDPAs) and workfare participants and collaborating employers (see Nativel, Sunley and Martin, 2003). The case-studies illustrate how the types and scales of problems encountered tend to vary in different types of labour market area, so that even if the basic policy thrust of the NDYP is common to all local areas of delivery, it has faced different specific conditions and challenges in different localities. Our aim is not to suggest that the programme has been completely without successes in depressed labour markets, nor, equally, that has

**Table 4.1** The case-study Units of Delivery

| Unit of Delivery | Delivery Model | ES Cluster Type | Estimated 18–24 Unemployment Rate 1997[a] | Per cent of Clients Entering Unsubsidised Jobs[b] | Percentage of NDYP Jobs Retained for 26 Weeks[c] |
|---|---|---|---|---|---|
| Birmingham | Joint venture partnership | G – Inner city high unemployment | 5.35 | 33.87 | 46.01 |
| Cambridge TTWA | Employment service led | C – Rural/urban tight labour market | 1.06 | 51.15 | 51.74 |
| Camden & North Islington | Consortium | G – Inner city high unemployment | 5.52 | 33.93 | 49.71 |
| Edinburgh &East/Mid Lothian | Employment service led | E – Urban tight labour market | 2.76 | 45.90 | 48.16 |
| North Tyneside | Private sector led | F – Urban high unemployment | 4.06 | 46.41 | 39.97 |

Notes:

[a] Estimated from Labour Force Survey Local Authority Database, Unemployed 18–24 year olds over population total 1997.

[b] Data relate to first seven cohorts, at April 2000, ES Core Performance Measure database.

[c] Percentage of those obtaining any job and not returning to JSA after 26 weeks, first seven cohorts at April 2000, ES Core Performance Measure database.

it been entirely problem-free in dynamic, higher employment growth labour markets. Although, there has been a distinct tendency for the policy to be more successful in the latter than in the former. The basic point is that no two local labour markets are the same, and policy outcomes are shaped in large by local circumstances.

This section briefly considers some of main problems encountered by the NDYP in more dynamic and buoyant labour markets. It suggests that its operation was characterised by 'creaming' as the most employable soon found work so that the programme increasingly had to work with the hardest to help and those with multiple disadvantages. Some delivery agents felt that the programme was insufficiently equipped to help such individuals and that, partly as a result, there tended to be serious problems in engaging and maintaining the support of local private employers in tighter local labour markets.

One of the features common to the programme recognised by officials in all of our case study areas was the tendency for 'creaming', or the tendency for the programme to be of most help to those with the most employability, in metaphorical terms, situated towards the front of the 'local queue for jobs'. We were frequently told that employers were primarily concerned with personal skills, motivation and willingness to learn, and that those young people with these qualities were far easier to help. As a New Deal district manager in the West Midlands explained:

> If someone comes on to the New Deal at the six-months stage and perhaps just lacks a bit of motivation, perhaps a bit of help with what they've got to do; if they look as though they could be placed quickly, then nine times out of ten, they will get a job. It's the ones who are more difficult who lack motivation, who come from generations of people who don't work.

In the more prosperous labour markets, the more employable individuals were soon placed and the main policy problem that soon emerged concerned individuals with multiple and serious obstacles to work. For example, in the buoyant Cambridge travel to work area, having moved the most employable people into employment, programme managers highlighted the fact that for the remaining 'hard core' of disaffected clients, raising employability is a more challenging task than mere work placement. A project manager for an option provider in Edinburgh also commented:

> We have low unemployment levels and there are so many jobs. Everywhere you look, on Princes' Street you see adverts 'We want staff'. So there's jobs available if people want them and are prepared to be flexible. So if people are unemployed still, there must be issues. We have a tight labour market and the client group is therefore the harder to help.

According to this manager, the compulsory element of NDYP had been more effective in 'digging out' people with problems. Interestingly, a New Deal project manager from Edinburgh told us that when they had visited Knowsley they had found that the client group there was much more mixed in that it included a range of participants with good skills who still could not get jobs, whereas in Edinburgh the client group was more homogenous and very disadvantaged. In both Cambridge and Edinburgh it was argued that the hard-to-help problem had become more serious over time as the more employable had been placed in work. According to the New Deal co-ordinator in Edinburgh, 'We're now dealing with youngsters who have social and behavioural problems, not an economic problem.' He described such individuals as having disorganised lifestyles who had no contact with job-holding and wage earning, and continued, 'For young people who have chaotic lifestyles a job is not the answer. What is missing in the Gateway is the intervention of voluntary and statutory bodies to help young people. They need to be stabilised.'

Several managers felt that the programme could not provide sufficiently flexible and individually tailored provision to help those with more severe problems. As a senior figure in the strategic partnership in Edinburgh put it,

> Because there are so many enormous personal and social deficits we need much more radical foundational thinking. New Deal worked with the top level of the unemployed who really just needed a job but the further down the pile you get, I don't think it can be the right mechanism.

The associated risk of labour recycling is identified by several Cambridge-based New Deal agents:

> We pushed them into this sausage factory and they're not coming out suited for our employers . . . It's the sausage machine. Feed it in there . . . Oh! That's a fitting thing, let's put it through again and see what happens (interview, Development Agency executive, Cambridge).
>
> If we're not very careful with this particular client group, we're just going to recycle them for evermore. New Deal was initially supposed to be about breaking that dependency culture. I actually don't think it will happen. The trend coming out nationally and the local performance target of moving people into employment are actually flagging that up (interview, local author-ity manager, Cambridge).

The evidence from tight labour markets also revealed problems regarding the involvement of employers. Initially it seemed a large number of employers had volunteered to be involved in the programme, expecting to see large numbers of potential employees:

One of the frustrations in Edinburgh is that we seem to have had many more companies volunteering to take part in the programme than it has been able to engage with, partly because Edinburgh is a high-employment area (senior figure in Strategic Partnership, Edinburgh).

A manager of the Employment Service's Eastern region commented:

I think a lot of employers at the time, particularly in buoyant labour markets, were seeing New Deal as a way of extending that pool of available labour because they were running out of other people to recruit. I think there was that initial high and then you go to the low of people saying, 'But these people aren't ready for a job,' and that's when we had to say to them, 'But that's the point.'

Employers who expressed a willingness to provide subsidised job options were often disappointed by the lack of referrals in these areas with a low supply of New Deal clients, and consequently many firms lost enthusiasm. The same problem was reported by managers in other areas and the New Deal marketing campaign was criticised for building up expectations that there was a lot of young people ready and waiting for jobs. As one district manager put it, as a result of the publicity drive, 'People thought that New Deal clients were going to come gold plated and with no rough edges. The reality is they don't.' Several employers expressed the view that anyone unemployed long term in such a labour market must either be completely lacking in basic skills or had chosen not to work.

The gap between employers' expectations and the character of the clients supplied through the programme was evident. Many employers complained of inappropriate referrals and were wary of NDYP clients. While some employers stated that they had lowered their demands in response to the shortage of locally available labour, others expressed the opinion that there was a chronic shortage of skilled labour and that contact with young people through the programme had reinforced this perception. Despite the skills shortage, several employers in Cambridge expressed reluctance in investing heavily in training young people, as they feared that these individuals would only be lost through high labour turnover. But there was little doubt that a buoyant local labour market helped the operation of the programme. A business manager at an Edinburgh job centre commented:

Now with the labour market in Edinburgh being as buoyant as it is, a lot of employers are prepared to take people on who perhaps they wouldn't have been prepared to take on two years ago because their expectations of peoples' skills would have been higher at that stage. Now they will consider people who perhaps don't have the level of skills. They'll be prepared to provide extra training, additional help with someone.

The presence of labour shortages in some sectors and places could also be highly beneficial. The Edinburgh Unit of Delivery, for example, established intensive sectoral Gateways that linked the introductory training of clients to opportunities in certain sectors, such as finance, where considerable numbers of vacancies existed.

## New Deals in Big Cities: Job Search and High Volumes

It was not only in buoyant local labour markets that specific problems of low employability, and distinctive difficulties in the operation and mechanics of the New Deal, could be found. Our case studies in London and Birmingham also revealed that in addition to generic issues, there were also peculiar problems encountered by the programme in large inner urban labour markets. While the NDYP here benefited to a certain degree from employers' demand for young unskilled labour, these benefits were constrained by serious problems of mismatch, some of which were geographical. In such large labour markets immobility and restricted job search became important issues as many clients proved reluctant to travel to distant and low-wage work opportunities. This lack of mobility was entangled in complex ways with the fact that large numbers of clients belonged to minority ethnic groups. There was also evidence that the high volumes of clients and the pressures on New Deal staff in such local areas meant that frontline staff were typically forced to compromise on the quality of the services they provided to clients. Furthermore, we note that high housing costs and the effects of the housing benefit system in some cities also acted as a disincentive to the acceptance of entry-level jobs.

Our interviews in the Camden and North Islington Unit of Delivery area revealed a distinctive local labour market that in some ways combined both features of high- and low-demand areas. Officials in Camden stated that the local programme did not suffer from a shortage of employment demand, because they had been able to link up with the West End job centre and hence with large organisations in London's West End, particularly head offices, and businesses in the hospitality and tourism industries. These vacancies supplemented the more restricted range of local vacancies in retail and supermarkets, administration and warehouses. In general the demand context in London was seen as favourable to the programme:

> If the time is not right now, it will never be. People are becoming so desperate now for good quality staff that they can invest in. They've never been so amenable. We're in a situation where employers need us. This is the sort of climate where the job is most rewarding (district manager, London).

Our local case studies also found repeated indications that the effective job-search areas of those involved in the programme were locally constrained. We know that higher-income groups tend to commute further (see Benito and Oswald, 1999) and the practical and psychological barriers to greater travel-to-work mobility among young people are considerable (Nathan, 2000). The cost and inconvenience of public transport act to shrink the real feasible travelling distance of many young workers and our interviews with young New Deal participants show ample evidence of an attachment and belonging to their respective local areas, which results in a low propensity to relocate for work and a low propensity to look for work outside their communities: this was particularly the case among less well-qualified groups. For various emotional reasons, such as the proximity of parents, friends or partners, many young people are reluctant to look for work outside their present travel-to-work area.

This problem was strongly expressed in our interviews in North London. Advisers expressed some frustration that despite the availability of subsidies for travel (specifically a travel card enabling low cost travel), young people were reluctant to travel to accept vacancies in neighbouring areas. In the view of one adviser this was partly a financial question:

> There's always been a question about the type of vacancies that are being supplied in this area to meet the demands and expectations of the clients. We could probably get more further afield but the question is: are the clients prepared to go further afield to areas where the type of work that they are looking for is more in supply? For example we have clients that are looking for work in hospitality. The vast majority of those vacancies are in the West End. But of course, the industry historically has been known as being fairly badly paid. If someone is going out from here to work in Knightsbridge at maybe £4.50 an hour, once they've worked that all out and they've done all their calculations, they'll find, 'Is it really worth it?' They've got to figure out about travel, about unsociable hours and getting back here – all of that comes into account.

Others argued that the basic problem was that for many young people long-term unemployment had created a mentality and routine that had narrow horizons, that they were not used to the idea of travelling to work and that a 45-minute journey was often a deterrent to taking a job. According to a district manager in Camden, compared to rural areas,

> Here the transport is fantastic but there are still people who won't step outside their borough, sometimes their housing estate. There is this huge hang up with crossing the river. The motivations for people not wanting to travel differ hugely and the only way of finding out what they are is to talk to individuals. Sometimes it's fear – it might be racially induced, if a person from a certain

community crosses the road, they might get beaten up – it might be fear because I don't know how to get around – and that is frighteningly common – sometimes it's territorial and I think the north–south thing comes in. Sometimes it's financial and people are concerned that they're not going to be able to make the books balance (interview, district manager).

In some local areas, ethnic and other social issues also act to deter mobility. In Birmingham, for example, one mentoring co-ordinator argued that ethnic tensions and racial fears were also partly responsible for the reluctance to travel to work. Some Asian women were reluctant to accept work that brought them into contact with men and

> It's such a big place. People say, 'I don't go over that side of town.' I was thinking, 'What is this thing about'? It's about how Birmingham as a city operates. People operate locally, so if you live in Handsworth you ain't going to Northfield 'cause that's a no-go area for black people.

A distinctive feature of both the London and Edinburgh studies was the degree to which the local interactions of the housing market and benefits system created a 'benefits trap'. NDYP participants were reluctant to accept relatively low-wage jobs because of high housing costs and more specifically for fear of losing their housing benefits. Even where benefits could be 'passported', the lag in establishing payments could be a deterrent to accepting work (see McQuaid and Lindsay, 2002). According to a senior adviser in Edinburgh, many young people had their own accommodation and were looking for a level of income that employers were not prepared to pay:

> If you look at Edinburgh as a city, it's had a strong emphasis on the retail trade as well so there are retail opportunities. But a lot of clients that I have spoken to in the past have issues concerning housing because housing costs in Edinburgh are very expensive...We're looking into housing costs, council tax benefits and their salary. We wouldn't want to advise somebody taking a job if they're gonna be worse off.

The problem in London was not so much a shortage of vacancies; rather it concerned the quality of the vacancies on offer and the disincentive effects of low salaries. Some advisers felt that some of the objections to working in the hospitality and retail sectors and some public sectors were exaggerated and based on myths, and several units had organised events in which clients could meet with employers to try to improve the reputations of these sectors. The unpopularity of work in some of the expanding service sectors is not confined to London. McQuaid and Lindsay (2002) also report that many unemployed people in Edinburgh have selective job-search strategies and are reluctant to move beyond traditional job roles because of their concerns about the insecurity and low disposable income of entry-level positions in the service sector, compared to the relative security of benefit payments. As we will see, the same issues concerning the quality of available jobs were strongly expressed in our more depressed local labour markets.

Officials in the Units of Delivery in both London and Birmingham were aware of the particular difficulties that they had in getting young people from ethnic minority groups into work. Again this was clearly a product of the interaction of both supply and demand-side factors. For instance, a manager in North London conceded that their equality of output statistics for ethnic minorities show a lack of parity with white groups, but explained this mainly in supply terms:

> The areas where we have discrepancies were for our ethnic minority clients and in many cases it's not surprising given that we have quite high levels of recent immigrants who have poor spoken English, who have housing to secure, who have some real adjustments to make when they moved from Kosovo to Central London. And it's not surprising that they're not moving into jobs in the same numbers as their white counterparts.

She added that the statistics were sometimes misleading because of the high propensity of Asian clients to go into higher education, but the Unit was looking to involve ethnic minority businesses in the programme in order to help those clients who would prefer to work within their own community. Another official argued that some black Afro-Caribbean clients did not have a good understanding of the way in which many employers expected them to speak or to dress. Other officials suggested that there were employers who continued to discriminate. For example, when discussing this issue, we were told that City of London firms were looking for 'the right person for the job' and were unwilling to compromise. An official in the Birmingham Enterprise Zone told us that many employers had a problem with ethnic minority groups as they assumed they had 'an attitude'. Such factors may partly explain why, despite the positive experiences of more employable individuals from ethnic minorities on the NDYP, there have been widespread problems with parity of employment attainment and disaffected attitudes among those who avoid or drop out of the programme (Fieldhouse et al., 2002a, 2002b).

What is also clear is that Units of Delivery in large cities with large volumes of clients had a number of particular difficulties. One of these was the difficulty of monitoring the destinations of young people in these areas.[1] More significantly, the large number of clients handled by personal advisers, particularly in the early stages of the programme, meant that advisers were unable to get to know, refer and support clients properly.

Further, high case loads appear to have had a negative impact on staff morale and turnover. As a district manger in North London explained:

> The smaller the job centre, the better you know your client. And the best job centres in my view are the ones where you've got a register of 600 to 800 and you can actually give a one-to-one service and have some credibility about it because you really do remember the person from the last time they came in. To contrast that with here, I have 30 per cent staff turnover. Many of the staff are relatively inexperienced. So getting the basics right so that they can provide a good service to clients is a challenge.

The problem of high turnover among New Deal personal advisers has been reported in both the NDYP and in other New Deals (see Atkinson et al., 2000). It has undoubtedly been aggravated by the stress caused by high caseloads for individual advisers. One adviser in North London stated that he had had a caseload of up to 80 people, which was unmanageable and 'like a conveyor belt'. Another complained that he was too busy reacting to client problems to be able to plan his work,

> The way I deal with it, I try to get rid of the person in front of me and see the next person. It could be a horrible problem that's gonna take an hour to sort out. We don't have someone to deputise and take over our caseloads. It's just whoever is here. We don't have control over our day.

The same adviser stated that visits to providers were difficult in practice, as they were 'totally bogged down in paperwork'. There were clear indications then that the pressure of work and targets on frontline staff were leading them to compromise some of the aims of the personalised caseworker approach. As Wright has argued, job centre staff respond to pressures 'by redefining what it is that they are aiming to achieve during their interactions with clients and in this way they re-create unemployment policy'(2002, p. 249).

## Expectation Gaps and Job Quality

We now turn to the problems facing the New Deal in distressed and structurally depressed local labour markets, typically in areas of pronounced manufacturing decline. Here the programme in many ways faced its most challenging and difficult terrain. Our case study of North Tyneside, as well as interviews in other depressed local labour markets in the West Midlands and Scotland, revealed severe combinations of policy problems. In such areas there were profound gaps between what many clients hoped to achieve, the types of vacancies seen as desirable, and the rewards and demands of existing vacancies. Delivery officials in such areas were often highly aware of the limited range of vacancies they could offer to their clients, reflecting the problems of weak and unbalanced employment growth typical of distressed local economies. The decline of heavy industries in these areas had also left an engrained problem of structural mismatch, and produced types of low employability that can be directly traced to relational contexts. One reflection of this was the many young clients who came from backgrounds where non-employment had become normalised and who consequently had very low employability; another was the reluctance, particularly among young males, to accept new opportunities in service sector work.

One of the most important problems facing the NDYP, which recurred in many of our interviews, was the mismatch between what many young people expected from a job and what employers expected from new recruits. Many agents involved in delivery felt that both groups often had unrealistic expectations, and that this expectations gap was important to understanding the weaker performance of the programme in some labour markets. It was especially severe in labour markets with lower demand, where most vacancies were seasonal part-time insecure and low-status service jobs. For example an official in North Tyneside expressed the view that

> There's a big problem in the North-East about young people's understanding of what is realistic when they first enter the labour market. I hear some New Dealers saying, 'I won't work for less than two-hundred quid a week.' They've got no qualifications, they've never worked in their lives, no employer is going to give them a job for £200 a week. That acts as a barrier.

A senior official in Wales explained the problem and their response in this way:

> What we are doing is helping with the first step. But people look at a job and say 'The money is only £140 a week' or 'The hours are too long' or whatever. What we're saying is not 'Take the job for the rest of your life.' We know employers preferences: if they've got a range of applicants, they take people with a recent history...It's the start of the journey, not the end, that we're looking at with New Deal.

But many officials also felt that employers had unrealistic expectations and at the same time were reluctant to offer training. A regional office manager in Birmingham told us that employers were often looking for somebody who could go straight into a job and perform, when in reality many young clients lacked basic skills:

> Certainly the training element is very important because we've traditionally had this mismatch between what employers want and what employees have to

offer. To be honest, it has been hard to get employers to accept that one of the things that they need to do in order to get the people they want is to be prepared to invest in training them.

Similarly an adviser manager in London commented that

> Barriers on the demand side involve employers not having a full understanding of the type of clients we are likely to refer. It's about making them aware. For them to have an Office package, IT skills might not be realistic for people who have been unemployed a long time. It's about making them more aware, more flexible, more determined to train clients.

The problem seemed especially acute in North Tyneside where New Deal officials were candid about the lack of variety and quality in the types of vacancies they could offer, and the need for more support to be offered to those employers accepting clients. One development manager commented:

> The biggest problem that I have with the New Deal is that it's a labour market initiative, but it's only looking at one side of the labour market. It only looks at supply; it only looks at what are the problems with the potential employees. It doesn't look at the demand side, at what support employers need if they're going to take on New Dealers. They need to know that if they are getting New Dealers they might have problems and they need to know how to deal with those problems. They need to understand why the subsidy is being paid and how that can be used; they need to understand what sort of training they can provide.

Demonstrating this gap in expectations, many employers complained of inappropriate referrals and little follow-up from the Employment Service. On the other hand, given the larger number of clients in depressed labour markets, employers could generally be more selective about the clients they chose to employ on a subsidised job and so eventually found suitable employees. But the interviews suggested that the imperative and difficulty of moving people into jobs meant that there was little evidence that job opportunities were being carefully selected and vetted (and some of the consequences of this are addressed in Chapter 5). Many of the employers who became involved in depressed labour markets were small 'lifestyle' businesses that could not afford to provide proper supervision and training. Policy managers in several case-study areas pointed out that there had been problems in encouraging employers to recognise that they had some responsibility to provide training.

Yet many New Deal managers claimed that in general there was no shortage of vacancies for programme clients to go into. For instance, an official in the West Midlands argued that there were plenty of sectors in which young people could find a job and that problems of mismatch were not serious:

> There are still engineering vacancies in the Black Country but we have a lot of retail vacancies, and tourism surprisingly enough. I mean tourism in the widest possible sense. We have quite a few hotels that have opened up, we have quite a lot of, dare we say, chains of pubs and restaurants.

She then continued by asserting that young people found work in a range of jobs:

> It's very varying: motor vehicles, hairdressing, care. We have a lot of care homes springing up, and always a lot of vacancies for care homes. But of course the amount of money paid, whilst it might be the minimum wage, it is not the expectations of the New Deal client and not everyone's cup of tea. Also, in childcare, it's very much a growing industry, if you like. Although we have a high proportion of vacancies, again the pay is low.

However, this quote also reveals that it is not sufficient that local job vacancies are plentiful: the nature of those jobs also matters. Many entry jobs in the hospitality, leisure and tourist sectors, for example, are low wage and insecure, and offer limited skills and career development. In fact, in some local labour markets it was clear that the lack of enthusiasm for entry-level vacancies meant that some young people had shown a preference repeatedly to go on the Full Time Education and Training option. An adviser in North London stated that many just seem to want to do courses rather than accepting entry-level jobs,

> After follow through they go back on to normal signing for a bit, and then they come back on to New Deal. Very often they come in and ask, 'Can I go on another course?' That's probably the most frustrating thing: you can get them on to all sorts of things, but that doesn't always get them any further forward.

At the same time others revealed that environmental and voluntary service options were seen by many as 'punishment'. In response to the preference for taking courses it was clear that 'work-first' priorities had been imposed:

> Initially it was very much led by the client. If they came in and said, 'I want to go on a course,' we'd arrange it for them right from day one; but we've been told now that we can't do that anymore – it's all to do with jobs. We must try and get them into jobs, unless there's barriers like very little English or special needs. We must get them into jobs first of all, and if that's not possible then look if a course is the right way forward for them, but only if a course is the right way forward for them, ... if it's going to lead to a job.

Finally, the gap between employers' expectations and perceived levels of employability was worsened in formerly manufacturing-dominated labour markets by structural shifts in the nature of labour demand. Employers often argued that in some ways the labour supply had failed, or was slow, to adjust to the decline of heavy manufacturing and 'metal-bashing industries', and that structural decline had left a legacy of severe problems in the local labour market. They suggested that in high long-term unemployment areas groups had adjusted their lifestyles to the money received on benefits and were in some ways comfortable, and the thought of leaving benefits was seen as far too risky and insecure. An economic development official in Tyneside argued that unemployment rather than education had become the norm in some areas and explained:

> If you look back in the 1960s when there were lots of jobs in the shipyards, coalmining, etc. the 18–19 year old who is on New Deal now would have been the sort of lad who could have gone just into the sort of job his dad does. They didn't need lots of qualifications, they didn't need to do anything at school, because there was a job there for them. That's the reality. The reality of the modern labour market is that it's much more competitive. A lot of people haven't made that adjustment.

According to a manager in the Black Country also:

> A lot of our clients have never worked, whose families have never worked, or have not worked for a long time, or whose families just went across the road to work. 'Well if my dad and granddad can go across the road to work, why can't I?' It's this kind of attitude. And to people in the Black Country, say in West Bromwich, Wolverhampton would be the other end of the earth. They don't want to go.

Whereas in the past, young men had been able to gain apprenticeships and factory work with few interpersonal and soft skills, several agencies pointed out that such opportunities were now scarce. They also suggested that the job expectations imparted by local communities and parents tended to deter young men from welcoming service-sector opportunities. A common theme in many of our interviews was that young people's employability was strongly shaped by their parents' experience, so that problems had become intergenerational. An official in a Leith job centre told us about the case of one young man on New Deal:

> We got him into employment. Because he had to get up early every morning, he woke up the dog; and as the dog barked, he woke up the parents and they were extremely angry about this. They actually put the boy out of the house and made him homeless because he was waking them up by merely getting up

and going to work in the morning. That's the type of issues we're dealing with. You can often persuade the youngsters that this is the right way for them; it's the pressure that they get from parents who have been unemployed for many years.

These intergenerational and family attitudes were also often found to constrain young people's willingness to travel to other labour markets and look for different types of work. An official in Midlothian explained to us that young people were reluctant to travel to Edinburgh to find work because of intellectual and emotional barriers. That is, they felt connected to a low-skill, low-status economy, well known to them and to their parents. They preferred to work within familiar social groups:

> It takes 25 minutes to get to the town centre, and jobs in the financial service sector are open to them, and in the hotel and catering industry they're available to them. It's only in their head that they aren't.

In our case studies of depressed labour markets, the programmes struggled to respond to some of the consequences of localised long-term unemployment. The gap between employers' expectations and perceived employability often had a geographical dimension, in that employers identified certain high-unemployment areas as characterised by cynical and pessimistic attitudes towards paid employment, and argued that the young people from such areas had been socialised into very poor job motivation. But in an indirect way such difficult local circumstances forced local delivery agents to be more innovative and to make use of large public sector employers, as we discuss in the next chapter. However, the complex combinations of mismatch, inappropriate expectations, low employability and poor job quality in more depressed labour markets have produced a number of problematic consequences.

## Workfare Recycling in Depressed Local Labour Markets

The first of these is that in more depressed local labour markets the programme's effectiveness in moving young people into sustainable jobs has been significantly lower and, partly as a result, there has been a recycling of long-term young people back into benefit claiming, thereby raising unemployment inflow rates. For example, Figure 4.1 shows the variations in the proportion of spells on New Deal that are restarts, averaged between January 1998 and September 2003. That is, it maps the degree to which different districts are witnessing the movement of young people right through the programme and back into unemployment and thence into the

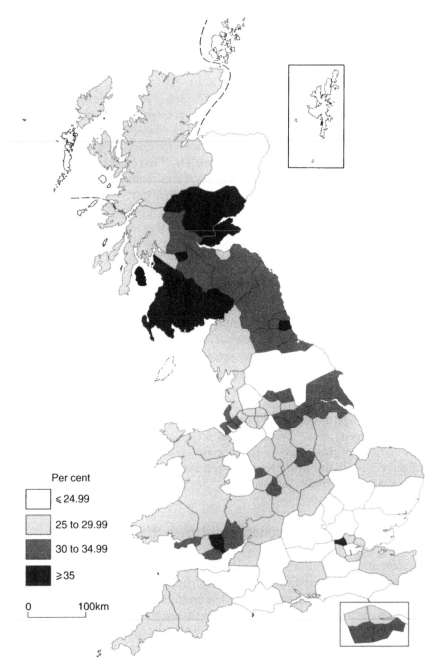

**Figure 4.1** Percentage of starts that are second and above, averaged between January 1998 and September 2003. Source: New Deal indicators, Department for Work and Pensions.

programme again for the second, third and even fourth time. The interpretation of these figures is not straightforward, as they are affected not only by the health of the local labour market but also by possible variations in youth mobility, by the extent to which disillusioned participants disengage from claiming, and by different management practices.[2] However, there are again strong signs that the local labour market context makes a difference as, while the proportion of second spells has risen everywhere, its rise has been very uneven. Consequently the lowest rates are mostly in rural areas in southern England, while the highest rates are in northern industrial cities, central Scotland and South Wales. The restart rate ranges from lows of 17.5 per cent in Surrey and 20.3 per cent in Hertfordshire to highs of 40.7 per cent in Fife, 38.5 per cent in Ayrshire Dumfries and Galloway, 35.7 per cent in Sunderland. 35.1 per cent in Glasgow, 34.7 per cent in Bradford.

What then are the causes of this geography of workfare recycling? It is undeniable that the compulsory element of the programme means that a significant number of young people in labour markets with fewer job opportunities feel that the programme 'pushes' and 'shoves' them into inappropriate jobs. This type of terminology is used very frequently by young people describing their meeting with their personal advisers:

> I didn't want to come here in the beginning because I felt that it was just a waste of time. He said, 'I've got to *push* you into this direction' (interview, North London, 7 November 2000; emphasis added).
>
> She kept trying to *push* me into doing secretarial work and temping and I don't want to do that. It's boring to me and I'm not interested in it (interview, North London, 17 November 2000, emphasis added).

Some of our interviewees argued that those Units of Delivery that took more risks by placing more young people into any available jobs during the Gateway tended to have lower retention rates, while those that were more cautious and patient tended to have lower job attainment rates but higher retention rates. Irrespective of the possible outcomes of such different practices, it is also clear that the low employability of some individuals is expressed in an inability to retain employment. This is typically a product of the interaction of multiple and persistent barriers rather than a single factor (Kellard, 2002). Some of our interviewees argued that many young people lacked basic employability and refused to accept that they had to start at the bottom with a poorly paid job. As a personal adviser in Camden explained:

> A lot of people get a job and sign off for a week, and very often they'll be back within a month or two. They get fed up with the job or they'll have an argument with the boss and leave. So I think they're not used to the idea of working. They can ignore the teachers and school and they think they can do the same at work, and get sacked.

Again however, there is undoubtedly an important demand-side dimension to the problem of low retention, in that many of the jobs available to clients were low quality, insecure and with no development prospects. For instance, an official in Birmingham expressed some frustration with employers' strategies:

> You're always going to have your packing jobs. A packing job is a packing job but if a company has in place a bit of training, the opportunity to progress from that, some sort of personal development, it can make that job not just a boring dead-end job: it can actually give that job a little bit of scope. But not everybody places emphasis on training their workforce.

The low wages of many entry-level jobs obviously tend to weaken attachment. For instance, Casebourne (2003) argues that, in the case of Sheffield, many lone parents have not retained the jobs acquired through the NDLP, not only because of childcare difficulties but also because many individuals, despite 'make-work-pay' tax credits, found that they had only succeeded in moving into 'working poverty'.

The variations in job attainment rates shown in Chapter 3 are perhaps of key importance to the recycling of young people, as relatively low levels of placement in employment tended to produce relatively high numbers going on to the post-option follow through stage and back into benefit claiming. On the basis of their national study of unemployment flows, Riley and Young (2000) argue that there is little indication that the slight rise in inflows to unemployment among the client group are the result of flows from employment back into unemployment. They argue that 'There is little evidence so far that the rise in inflows is due to early termination of jobs' (p. 20). They suggest there is no evidence of the New Deal inducing an increased rate of movement among young people from jobs to non-employment, or, to put it another way, that the jobs obtained via the programme do not differ from the normal level of job sustainability in the economy. The rise in inflows, in this view, would be due to those returning to unemployment after having participated in the training, voluntary sector and environmental programme options rather than from the job option (Riley and Young, 2000). This is highly plausible, both because the more employable young people will tend to enter the subsidised job option more quickly, and because the training and employment options tend to be used as a 'pressure release valve', or a soft option, in the most depressed labour markets (Martin, Nativel and Sunley, 2001). However, we have seen that at a local scale there are significant differences in job retention rates (for 26 weeks or more) across different localities (Chapter 3), and this may be entirely in accord with the 'normal', non-programme levels of job sustainability in these labour markets (after all, the normal level of job sustainability for the whole economy is only an abstract average of varying local rates). For instance, a local authority representative in the north-east comments that

> The main underlying factor in my mind is unemployment, because the labour market in the north-east doesn't create as much employment as in other parts of the country. The employment picture is characterised by quite low-skilled, low-paid, temporary, casual, part-time employment and there aren't the well-paid craft and manual type of occupations like there used to be. Often you find that people are being directed towards jobs that they don't really want and at the end of the day they don't stick them. New Deal is essentially about getting the benefit bill down, it's not about encouraging young people to train maybe for a year or two years to get a good sustainable job which is what it should be about. It's about a quick sort of fix. This is one of the key failures of New Deal, and it is a big impediment in the north-east where it is quite easy for advisers to place clients in low-skilled temporary jobs, but they're not addressing the problem; they're not increasing employability, they're not increasing the basic skills and qualifications (interview, development manager).

Part-time employment (with a working week of under 20 hours) was more frequent in London than in the other UoDs. This may be partly explained by the interaction of wages with the operation of the housing benefit system. The housing market in London seems to prevent young people from considering a wider range of opportunities, since interviews there suggested that the loss of housing benefits is perceived as a disincentive to taking full-time employment. By allowing a large proportion of young people to work on a part-time basis, the local New Deal agencies in Islington and Camden have used their discretionary powers to 'bend the rules' of the national framework within which they operate. Of course, because of its more precarious status, part-time work in the subsidised employment option may mean a greater exposure to the risk of workfare recycling.

Irrespective of its causes, it is important to consider whether high levels of workfare recycling are compatible with the New Deal's aim of raising 'employability' and helping participants to improve their prospects of remaining in sustained jobs. The Minister for Employment has recently argued that

> What the New Deal aims to achieve is a radical improvement in the employability of young people, giving them the necessary skills which in turn become personal assets, not just to get a job but to stay in work for the rest of their lives. There is also good evidence to show that people who do return to unemployment are leaving as least as quickly as short-term unemployed people. Even people who have been on the New Deal, who lose their jobs who return to unemployment, are less disadvantaged than they would have been had they not had the benefit of the New Deal Programme (Education and Employment Committee, 2001a, p. 182).

And some senior frontline staff share this optimistic view. For instance a more senior manager in London argued:

> One of the issues that we were first concerned with was that New Deal would look like a revolving door. People would come back and it would look the same. The reality is that it doesn't, because if people have been through New Deal the one time, when they come back they are re-entering it with a different set of skills, therefore with a different set of needs … So for many people they're concentrating on key and basic skills the first time round; they're coming back for more vocational job-focused effort the second time round (senior manager, Camden and Islington).

The Government's Response to the Education and Employment Committee's 2001 report on New Deal adds that young people often try out different experiences in the labour market while learning where their long-term goals lie (Education and Employment Committee, 2001b). But there are several problems with this view. One is that it mainly refers to those young people who attain jobs that, for whatever reason, do not last, and does not address those participants who go through the other New Deal options and do not find a job. As we have seen, it may well be that a greater part of workfare recycling is due to people going through these non-job options and then back into benefit claiming. The monitoring of the destinations of young people leaving the education and training, voluntary sector and environmental options has been almost non-existent. In theory, New Deal option providers have an interest in moving their clients into employment upon completion of a course or placement, as this positive outcome enhances their performance profile and thereby gains them more clients and resources. However, interviews with New Deal providers in our case-study UoDs revealed that the tracking of participants' destinations is uncommon, given the awareness that the New Deal is only a small proportion of their business and that it may well not be a permanent feature of labour market or welfare policy. Providers were only recently offered a cash incentive for 'job-outcomes', but it is debatable whether this is sufficient to provide the necessary resources to enable education and training providers to liaise closely with employers and thus facilitate the transition of participants into employment.

Second, it was clear from our interviews that recycling of clients was frustrating and demoralising to New Deal personal advisers themselves and that it therefore jeopardised one of the most successful features of the programme: the individual case relationship. As a manager in London explained:

> We do have a minority of young clients who refuse help or they can't keep in sustained employment so we're constantly seeing them coming back. It can be

quite demoralising when you are seeing the same people coming back, especially when we are trying to do our best. We've done quite a lot of work with them but despite that they are repeatedly coming back to us (NDYP manager Camden Islington).

Third, and more fundamentally, there are clear signs that low job retention and short-term benefit churning damage individuals' job prospects and employability (Kellard, 2002; Nathan, 2000; Peck and Theodore, 2000b). It is hard to deny that workfare recycling has a detrimental effect on young peoples' motivation, confidence and attitudes to work, especially when it does not involve any experience of unsubsidised employment. These are crucial components of 'work readiness' and employability so that the potential damage caused by workfare recycling needs to be treated more seriously. As a regeneration manager in Scotland argued, a more human capital approach is needed, since:

> Intensive things about changing people's attitude to being in a workplace do not happen through one-off interventions or a series of disjointed activities across the Gateway and filling them into a job from which they will fall out again.
>
> A monkey can place people into jobs when the labour market is climbing as it is when we have practically full employment. Anybody can place people into jobs. What they can't do is help those individuals to have sustainable employment because they've got the personal skills, the assertiveness, the self-belief to move within a changing labour market.

At the same time, it is being claimed that the programme is proving effective in reducing wage pressure in those regions and areas where is has been most extensively applied (Riley and Young, 2000). But this appears to limit and circumscribe the employability aim. If the New Deal is indeed having greatest downward pressure on wages in areas where it has the largest number of participants, then this could also tend to lower young peoples' attachment to their jobs and also their motivation and commitment to work (Sargeant and Whitely, 2000; Peck and Theodore, 2000b). Thus, in depressed labour markets there may well be a contradiction between welfare-to-work strategies and making-work-pay (Alcock et al., 2003; Gray, 2002). More generally, recent research on low pay in Britain has found considerable persistence and evidence of a cycle of 'low pay and no pay', rather than low-paid jobs acting as stepping stones to higher paid ones (Stewart, 1999). Perhaps more economists should recognise that if intense labour market flexibility can corrode the commitment, self-responsibility and work ethic of those in ordinary occupations (see Sennett, 1998), it is even more likely to be corrosive for those moving in and out of low-paid insecure employment.

## Moves to Inactivity

But the problem of churning and recycling is not the only important symptom of the degree of dysfunction of workfare-style policies in depressed labour markets. As we argued in Chapter 2, there is now a large problem of hidden unemployment in the UK, in that inactivity among men has risen substantially since the 1970s and there are now 2.87 million men of working age who are inactive (15.5 per cent of the 16–64 age group) (Alcock et al., 2003). The total number of adults of working age not looking for work has now reached a record high of 7.85 million. As we saw in Chapter 2, there is also a distinct geography to inactivity in that it is highest in areas of long-term claimant unemployment and strongly skewed to traditional industrial areas in northern England, central Scotland and South Wales, with high rates also in parts of inner London (Beatty and Fothergill, 2003; Webster, 2003). Furthermore, around 70 per cent of inactive men aged between 25 and 54 report themselves as sick or disabled (Nickell, 2004). As we noted in Chapter 2, the number of male incapacity benefit claimants has grown from about 400,000 in 1981 to 1.25 million in 2000 (Figure 2.5). The total count of people who are out of work and on sickness and disability benefits has increased to 2.7 million. The rates of claiming sickness-related benefits are also highest in places such as South Wales, Merseyside, Manchester, South Yorkshire, North-Eastern England and Clydeside (Figure 2.5; Beatty and Fothergill, 2003).What is less well known is that in depressed local markets in such areas, the New Deals may have contributed to, and suffered from, this flow into inactivity.

There is some evidence that in areas where there are fewer higher-quality jobs available, and fewer opportunities to find suitable work, the pressure applied on participants in the New Deal may be encouraging those clients with forms of ill health and disability to move on to other benefits, including incapacity benefit. Many of the jobs that used to provide work for individuals with ill health have disappeared as the demand for low-skilled workers has weakened, and as the labour market has become more flexible and more lightly regulated, so that there has also been a process of 'health selection' (Easterlow and Smith, 2003; Nickell, 2004). Those at the back of the jobs queue in depressed local labour markets have frequently become disillusioned with and detached from job search and have moved from means tested benefits on to more generous sickness related benefits (Beatty and Fothergill, 2003). At the same time, it may be long-term unemployment itself that is causing some to become ill (this is one of the interrelationships highlighted as compounding the unemployment problem in depressed local labour markets highlighted in Figure 2.10 in Chapter 2), and it may be that the personal attention under New Deal is recognising these complaints and moving people on to appropriate benefits.

Data on immediate destinations of those leaving the NDYP and the New Deal 25 Plus show that by the end of December 2003 over 160,000 people had left these two schemes for other benefits. There are strong geographical variations in the extent of this movement. Up to the end of 2003, the rates in the lowest areas indicated that only 7–8 per cent of leavers had left for other benefits. The lowest rates were in parts of London, in Surrey and in Cornwall. The highest rates tend to be in industrial urban areas where over 16 per cent of NDYP leavers have moved into other benefits. The highest rates were in Glasgow (16.8 per cent), Bridgend and Rhondda (16.8), Sunderland (16.2) and Durham County (16) (see Table 4.2). It is quite clear that even young people in depressed, formerly industrial labour markets are considerably more likely to transfer on to other benefits.

Not surprisingly, this pattern of flows into inactivity is even more marked for the New Deal 25 Plus. Not only does this New Deal show a high rate for returning to JSA (in many areas this reaches 25–30 per cent of leavers), but it also shows a high rate of leavers moving into other benefits. Indeed by the end of 2003 a total of 43,210 people had left the enhanced scheme for other benefits, and 78,040 had returned to JSA compared to only 63,810 who

**Table 4.2**  Percentages of NDYP leavers going on to other benefits, by selected JCP district to December 2003

| Districts with Lowest % Rates | | Districts with Highest % Rates | |
|---|---|---|---|
| South London | 7.2 | Glasgow | 16.8 |
| North London | 7.5 | Bridgend Rhondda, Cynon and Taff | 16.8 |
| London Central North-East | 7.8 | Sunderland | 16.3 |
| London East | 8.3 | Durham County | 16.0 |
| Surrey | 8.3 | Renfrewshire, Inverclyde, Argyll | 15.7 |
| London Central South | 8.3 | Lanarkshire | 15.6 |
| London South-East | 8.5 | Gateshead and South Tyneside | 15.1 |
| London North-West | 8.6 | Wakefield | 14.9 |
| London Central North-West | 8.8 | Cumbria | 14.8 |
| West London | 8.8 | Doncaster | 14.8 |
| Cornwall | 8.8 | Eastern Valleys | 14.7 |
| Berkshire | 9 | Tees Valley | 14.6 |
| Highlands Western Isles | 9.1 | Swansea Bay | 14.3 |
| Essex | 9.2 | Tayside | 14.3 |

Source: DWP (http://www.dwp.gov.uk/asd/ndyp.asp).

had left for unsubsidised jobs (Department for Work and Pensions, 2004). In terms of the percentage of leavers, this flow into inactivity is lowest in Cambridgeshire, South London, Suffolk, Devon and Cornwall at 12–13 per cent (see Table 4.3). In contrast, the highest rates of over 20 per cent are in northern industrial cities. The highest rates of about 25 per cent are in Doncaster and Knowsley and Sefton.

On the basis of such figures David Willets, the Shadow Work and Pensions Secretary argued that

> This suggests that the New Deal is moving people off official unemployment into hidden unemployment. They are clearly parking people on disability benefit in areas of high unemployment. Once you're on these benefits you're even further removed from the labour market than before (Turner, 2004, p. 3).

**Table 4.3** Percentages of enhanced New Deal 25 Plus leaving to other benefits, by JCP district, to December 2003

| *Districts with Lowest % Rates* | | *Districts with Highest % Rates* | |
| --- | --- | --- | --- |
| Highlands and Western Isles | 11.6 | Doncaster | 25 |
| Cornwall | 12.3 | Knowsley and Sefton | 24.7 |
| Surrey | 12.5 | Glasgow | 23.5 |
| North-West Wales Powys | 12.7 | Lanarkshire | 22.5 |
| Buckinghamshire and Oxfordshire | 13.0 | Wirral | 21.8 |
| London North-West | 13.1 | Salford and Trafford | 21.6 |
| London South East | 13.2 | Newcastle and North Tyneside | 21.2 |
| Cambridgeshire | 13.6 | Bradford | 21.2 |
| South London | 13.4 | Gateshead and South Tyneside | 20.9 |
| Devon | 13.6 | Manchester | 20.6 |
| West London | 13.6 | Oldham and Rochdale | 20.3 |
| Grampian Moray Orkney and Shetland | 13.7 | Stockport | 20.2 |
| London Central South | 13.8 | Renfrewshire, Inverclyde, Argyll | 20.1 |
| Suffolk | 14 | Bolton and Bury | 20 |

Source: DWP (http://www.dwp.gov.uk/asd/ndyp.asp).

In response to these problems, John Philpott one of the important architects of welfare-to-work in the UK, suggests that everyone on benefit should be required to seek work and that employers should be given larger subsidies to take on the disabled and sick (Turner, 2004, p. 3). However, this response still seems to take inadequate account of the geography of the problem. As the experience of the NDYP suggests, wage subsidies are unlikely to be sufficient to compensate for low levels of demand in the most depressed labour markets, where the sick and disabled are concentrated, and are unlikely to be able to improve the capacity of labour markets significantly in these localities to absorb the tens of thousands of individuals now inactive.

## Conclusions

This chapter has tried to provide a picture of some of the main employability and operational problems faced by the NDYP in different local labour markets. Of course, there are common themes and factors that recur in all our case-study areas. Nevertheless, the balance of supply-side and demand-side factors in causing low employability vary in different local contexts. The distinctive group of individuals encountered in buoyant labour markets with multiple severe disadvantages and barriers to work is a clear indication of this. The problem of geographically restricted job search in larger cities is another. In more depressed labour markets the low quality and restricted range of many vacancies has come into conflict with clients' expectations and their hopes to find secure and well-paid work in traditional sectors. The problems here are as much to do with the quality of vacancies as with unrealistic expectations. But what is clear is that the dysfunctional and damaging outcomes of these differences have included higher rates of recycling through the programme and higher rates of movement on to non-means tested benefits in depressed local labour markets.

If low employability has both demand- and supply-side determinants that are locally variable in this manner, then this has significant implications for the design of workfare programmes. The obstacles to greater labour market participation vary across individuals and localities, so that workfare programmes need to provide policies and options that are differentiated and responsive on both these scales. As Alcock et al. (2003) rightly argue, the New Deals have to be both person sensitive and context sensitive so that

> The balance of resources deployed in different parts of the country needs to reflect the particular local incidence of obstacles to labour market participation – which means an emphasis on economic development and job creation

in the areas where labour demand is weak, and on training and retraining where skill shortages are the main obstacle (p. 267).

In the following chapters we look at the employers' involvement in the programme and the degree of local flexibility in its design and delivery to assess how far such a locally sensitive programme has actually been delivered.

# Chapter Five

# A Geography of Mismatch? Employers, Jobs and Training

## Introducing Gatekeepers' Tales

Throughout this text we have placed substantial emphasis on the importance of differences between local labour markets. However, in doing so we do not wish to suggest that local labour markets are simple, internally homogenous entities. In truth they are internally differentiated and segmented markets, made up of complex institutional and social relationships (Peck, 1996; Martin, 2000; Reimer, 2003). Employers' labour practices and strategies are obviously a key element of this differentiation and complexity (Peck and Theodore, 1998). Employers are the 'gatekeepers' who control entry into jobs and set the conditions regarding the length of time (or tenure) of employment (Hasluck, 1999). Moreover, firms are clearly not autonomous; they are subjected to a range of internal and external influences that often intersect with one another in complex ways, and there is an important debate about the relative significance of local labour market influences on individual firms and employers (Rubery and Wilkinson, 1994; Grimshaw et al., 2001; Rubery et al., 2002).

Employers have certainly been central to the local operation of the NDYP and differences in the character of their engagement have been evident both between and within local labour markets. The types and levels of vacancies, wages rates and conditions have been a key influence on the local working of the programme. The fact that it has proved more successful in buoyant labour markets appears to vindicate Peck's (1999, p. 362) warning that the New Deal would 'subsidise pre-existing private sector vacancies' and 'ossify extant patterns of labour market inequality'. While even depressed local labour markets are internally heterogeneous, and often contain small niches of sustainable and rewarding employment opportunities, our results suggest that these are firmly in the minority, and contingent upon employers' relative exposure to the intensity of market competition.

This chapter unravels the intricate motivations and pressures that drive employers' attitudes and practices in respect of the NDYP. Drawing on survey work and in-depth interviews with employers and young people, it first examines how far employer characteristics and the opportunities offered to New Deal recruits vary significantly and consistently between different types of local labour markets. It does so by analysing the vacancies and training provided under the subsidised employment option, as well as the significance of financial incentives and disincentives. Special attention is paid to the role of the wage subsidy and its influence on employer behaviour.

The chapter then turns to the determinants of employers' involvement and non-involvement in New Deal for Young People and considers how involvement relates to firm-specific characteristics such as prevailing managerial cultures, workplace rules, and social ties. We develop a typology of employer behaviour and motivations, and discuss the interactions of this typology with the local labour market context. It is argued that advantageous initial labour market conditions and positive employer practices mutually reinforce each other to create a relatively broader range of employment and educational opportunities for young people (a positive sum game), while a strong degree of employer commitment in the context of unfavourable labour market conditions has a slightly mitigating (but not necessarily long-lasting) effect on labour market disadvantage.

## Variations in Job Opportunities and Recruitment

A comprehensive body of qualitative research and large-scale surveys has been carried out as part of the national evaluation of the New Deal for Young People in order to shed light on the employment and training opportunities resulting from the programme (see Snape, 1998; Tavistock Institute, 1999; Elam and Snape, 2000; Hales et al., 2000; Bryson, Knight and White, 2000; Bonjour et al., 2001). The vast majority of these commissioned reports acknowledged the importance of geography on employer recruitment via the New Deal. For example, the Tavistock Institute argues that 'different types of employers in terms of size, sector and *local affiliation* have different motivations for becoming involved in New Deal and offer different job and training opportunities to New Deal clients' (p. ii, our emphasis) and that the extent and degree of delivery problems affecting employers 'seems to relate to local labour market circumstances' (p. 41). A further report finds that 'employers seem more willing to recruit from New Deal when labour market conditions are tight or when they have a strong link to the local community in which they are located' (Hasluck,

2000, p. 50). Such statements hint at the importance of geography, without offering spatially disaggregated detailed findings. Nevertheless this literature provides a useful basis against which to compare the results of our research.

Existing research has documented the types of vacancies offered in each broad sector of activity against specific categories of employers and shown that the predominant occupational category is that of office juniors (clerical and secretarial assistants), as these are common in all sectors. Together with manual craft occupations, this category represents about half of all the vacancies in the subsidised employment option (see Table 4.2; Hales et al., 2000; Bryson, Knight and White, 2000; see also Atkinson, 1999, p. 23). Compared to the New Deal for the Long Term Unemployed (NDLTU) the amount of placements in clerical and secretarial occupations was higher in NDYP, while those in unskilled vacancies were lower (Hales et al., 2000, p. 59). This difference is likely to be due to the requirement to provide positions with some training component in the NDYP in contrast to the NDLTU, which some employers perceived as being to the disadvantage of older unemployed people.

In addition, evaluation reports have identified a number of unsurprising gender differentials. In contrast to the pattern of male recruitment, females recruited to subsidised jobs were almost exclusively employed in either clerical and secretarial or in personal and protective service jobs such as nursing assistants. Young males were more likely to fill manual semi-skilled and craft related vacancies. There appeared to be no significant gender differentials concerning retail vacancies or jobs offering higher levels of responsibility and autonomy (see also Hales et al., 2000, p. 60).

Table 5.1 shows the main types of vacancies organised by different types of business, as found in our survey. We first analysed the results by arranging the vacancies offered by employers into nine Standard Occupational Categories (SOCs) and by examining their occurrence across our case-study areas.[1] The last three categories, representing the higher end of the occupational scale (managerial, technical and professional), were represented as follows: 4.5 per cent of young people were recruited to this type of occupation in our North Tyneside UoD sample, 10 per cent in Birmingham, 13 per cent in North London, 16.5 per cent in Edinburgh, and 27 per cent in Cambridge. It is thus striking that the more buoyant labour markets of Edinburgh and Cambridge provided more jobs with higher levels of responsibility and skill. Skilled and semi-skilled occupations were the most representative occupational category for Birmingham (30 per cent), reflecting the relatively high involvement of employers from the manufacturing sector. This occupational category was also well represented in Cambridge and Edinburgh (above 25 per cent), but more insignificant in North Tyneside and in North London. Finally, the exceptionally high

**Table 5.1** Vacancies offered through the subsidised employment option

| Sectors of Activity | Type of Establishment or Business[a] | Type of Placement / Job Status[b] |
| --- | --- | --- |
| Public sector | Hospital | Nursing cadet; clerical assistant |
| | Government Department | Receptionist; clerical assistant |
| | Transport Authority | Customer assistant |
| | Police | Clerical assistant |
| Trade and manufacturing | Building and construction | Trainee carpenter; plasterer, joiner |
| | Car manufacturing | Trainee engineer |
| | Manufacturer (granite products) | Trainee polisher |
| | Manufacturing (metal products) | Power press operator; assembly worker; trainee fabricator |
| | Food manufacturing | Trainee sugar boiler; administrative assistant; receptionist; trainee photographer; gardener |
| | International trade | Office junior |
| | Fashion designer | Trainee dressmaker |
| Personal and protective services | Leisure centre | Fitness instructor; trainee outdoor pursuit instructor |
| | Complementary therapy centre | Receptionist/general assistant |
| | Tanning centre | Assistant manager |
| | Nursing home | Care worker; kitchen assistant |
| | Private hire | Driver |
| | Solicitor | Administrative assistant |
| | Gardening services | Trainee gardener |
| | Cemetery | Gravedigger; trainee letter cutter |
| Other services | Gas company | Call-centre agent |
| | Media company (film and television, publishing) | Production assistant; web designer, administrative assistant |

| Category | Type of establishment or business | Type of placement |
| --- | --- | --- |
| | Installation and design of telecommunication equipment | Junior engineer |
| | Consultant / financial institution | Receptionist; office junior |
| | Recycling, refuse collection | Paper processor |
| | Water treatment and softening | Trainee service engineer |
| | Drama School | Teacher/general assistant |
| | Caravan repair services | Caravan technician |
| Retail, catering, hospitality | Bar / Restaurant / Pub | Chef; assistant manager |
| | Catering company | Kitchen porter, waiter/waitress |
| | Hotel | Decorator; laundry attendant |
| | Travel agency | Administrative assistant |
| | Various retail outlets: books, food, furniture, music, shoes, pets, records, electronic/fishing/ surfing/art equipment, jewellery | Sales assistant; warehouse assistant; workshop technician |
| Community businesses | Cash and Carry | Warehouse assistant |
| | Local economic development | Administrative assistant |
| | Counselling and legal aid | Administrative assistant |
| | Housing insulation services | Cavity wall insulator |
| | Football club | Football coach |
| | Charity project (music and media) | Trainee in computer animation |

Notes:

[a] Note that the column lists a *category* of establishment or business. Each category may be representative of several interviews conducted with similar employers. Hence several types of occupations have been listed in the third column. For example, three hospitals (each in a different UoDs) were surveyed. Each had used New Deal for different purposes.

[b] This column reflects a *type of placement* offered to young people on the subsidised employment option. In several instances, a given type of occupation may have been offered to more than one recruit.

prevalence of occupations in the 'sales' and 'protective and personal services' (i.e. jobs as shops assistants or care assistants) in North Tyneside
(over 50 per cent) was striking. Birmingham displayed the lowest level of
employment in this category (20 per cent), possibly once again, a reflection
of its higher representation of vacancies in factories and other industrial
outlets. The three other UoDs displayed similar levels of New Deal subsidised jobs in this occupational sector (approximately 30 per cent in the
three cases). Our findings based on an average for five urban UoDs differ
from the national commissioned surveys regarding several occupational
categories, especially sales vacancies which were more commonly reported
by the employers we surveyed (see Table 5.2).

Of course, broad occupational categories do not determine actual job
content and future career prospects, which are particularly crucial in the
case of jobs at entry level. Subjective aspects such as job satisfaction and
self-esteem are important factors to take into account when analysing the
working of labour markets (Fevre, 2000), and the quality of New Deal
vacancies appears to have had a strong influence on the tendency of young
people to stay in work. For example, a shop assistant interviewed in Edinburgh was offered the prospect of becoming the manager in the near future,
while another young person in the same Unit of Delivery had been made a

**Table 5.2**  Occupational breakdown of subsidised job vacancies offered under the NDYP

| Standard Occupational Classification (SOC) | National Survey of Employers[a] | National Survey of Participants[b] | Authors' survey (Average of the Five Case Study UoDs) |
|---|---|---|---|
| Managers and administrators | 1 | 4 | 3.6 |
| Professionals | 1 | 1 | 1.2 |
| Associate professionals and technicians | 4 | 6 | 9.3 |
| Clerical and secretarial | 27 | 17 | 20.9 |
| Craft and related skilled manual | 25 | 23 | 17.8 |
| Personal and protective services | 12 | 11 | 15.2 |
| Sales | 5 | 11 | 17.3 |
| Operative and assembly | 12 | 9 | 4.3 |
| Routine unskilled manual | 12 | 19 | 10.4 |

Sources: [a] Hales et al. (2000, p. 59); [b] Bryson, Knight and White (2000, p. 93); Authors'
Survey.

partner of the business. Furthermore, in a number of instances, shop owners had used New Deal to recruit young staff with the skills required to help them develop new activities such as websites and e-commerce, and hence contribute to the expansion of the business. More rarely, New Deal was used to help recruits gain creative skills in media and the arts. In such instances, young people felt empowered and expressed the opinion that they were offered a 'fair deal'. They were less likely to associate job quality with high earnings but rather with the opportunity to develop their talent. These types of employment contracts were slightly more common in Edinburgh, London and Cambridge where leisure and cultural activities are relatively better represented; young people had directly approached their employer, who tended to be part of their social network to ask them to employ them via New Deal. Employers in the cultural and creative industries sector commented that they often had to turn down young people offering to work for them.

For private companies that had recruited several New Dealers, examples of a dual internal labour market were not unusual. For example, a London-based employer reported having recruited two sets of New Dealers: those he described as working on the 'retail side' and those employed on the 'Internet side'. He admitted that those working on the Internet side had scarce skills and consequently proved to have better employment prospects than the second set of young people who found themselves in a less secure peripheral position.

The possibility of progression towards a career and professional status is a central issue as it can drastically improve the long-term employment prospects of young people and reduce the risk of recycling of individuals through the New Deal. It has been illustrated elsewhere that young people associated long-term opportunities with the initial prospect of obtaining a 'good job' (Ritchie, 2000; Finn, 2003). Moreover, job dissatisfaction among young people proved to be high, with 79 per cent of those in the subsidised employment option stating that they would have preferred to work in a different job (Bryson, Knight and White, 2000, p. 88). Once again, the nature of local labour market appears to have been decisive, as an ES manager argues:

> The reality of this area is that we have a lot of hotel vacancies on our books at any one time and a lot of clients saying, 'I don't want to work there: there's no career path, there aren't any benefits, the pay is crap, the hours are lousy.' It's the employer's job primarily to make their job attractive (ES manager, London).

However, when reflecting upon the nature of the employment they offered to young people, employers – especially those recruiting for manual unskilled and semi-skilled vacancies – tended to express the view that these were not conducive to an attractive 'career'. On numerous occasions, employers described either the rate of pay or the nature of the vacancy in negative terms:

> You're not looking for commitment for life for the kind of wages that we're offering (SME, Edinburgh).
> they don't get a dreamy income (SME, North London).
> it's not the most glamorous sort of job in the world (SME, Edinburgh).
> It is not their career move to come and work in a press shop or in an assembly (SME, Birmingham).
> It's not the most pleasant of jobs. I would suggest that maybe 20 per cent of the people who've left, it's because they just can't handle it. I don't knock them for it. It's totally understandable. It's not an easy job (SME, Edinburgh).

A major tension appeared in that, despite this type of 'realism', employers were expecting young people to be enthusiastic, reliable, motivated, 'willing to work' and to 'do a good job'. In the 'employability' agenda, these types of 'soft' attitudinal skills or assets tend to be those most commonly sought by employers (Snape, 1998, p. 34; Hillage and Pollard, 1998; McQuaid and Lindsay, 2002). In interviews, employers commonly expressed the view that the acquired routines (or 'bad habits') of unemployed young people were incompatible with the discipline required in the workplace:

> What makes them unemployable is if they are unable to fulfil the demands of a working environment. So if they are unable to turn up at 8.30 am regularly, they become unemployable. And one of the problems with unemployment is that you drop into that routine of going to bed late and getting up late (SME, Cambridge).

Employment Service and other public sector officials often argued that employers hold unrealistic expectations of employability assets, in part fuelled by the early marketing campaign. Most young unemployed people entered the labour market with a significant psychological and social 'baggage' and little experience of the world of work. Nationally, it has been established that 80 per cent of New Dealers had at least one of four markers of disadvantage: living in social rented accommodation, no qualifications, no prior work experience, or suffering from a health or disability problem (Bryson, Knight and White, 2000). Furthermore, it was striking that previous work experience, academic qualifications, technical skills, and intelligence were rarely mentioned as expected assets, these being more often valued by employers recruiting for higher white collar (professional or technical) occupations:

> I'm looking for skills and flexibility, a range of things, an intellectual curiosity, being able to work supervised and unsupervised, showing enthusiasm, dedication, drive, intelligence – all those things (SME, North Tyneside).

But as we have seen, these types of employment vacancies have been relatively scarce, particularly in depressed labour markets.

Finally, one important facet of employability is the presentation and physical appearance of New Deal clients, which proved a particularly decisive recruitment criterion for companies operating in high customer-contact environments, such as the catering and retail sectors. The quotations below from two Cambridge-based employers explicitly refer to this issue:

> Our customer base is a bit upmarket. Therefore I look for people who are intelligent, well dressed. I have to say, if someone came in with tattoos and ear-rings and studs, I wouldn't employ them. I don't care which sex they are, because of the sort of customer base we have, they need to be able to conduct conversation with all manner of people at the right level (SME, Cambridge).
>
> These two are pretty clueless as to how they are expected to dress until they're told. I mean, basically eyebrow studs or nose rings, but they don't always realise that, so they'll turn up for a job interview wearing a nose stud or an eyebrow ring. If they turn up in a shop wearing a nose stud or an eyebrow ring, they won't even get an interview. If they turn up here wearing it, they will get an interview, but they will be told in no uncertain terms that it's not acceptable to wear while you're working here (SME, Cambridge).

During the New Deal Gateway, candidates are entitled to a £200 subsidy to buy smart clothes, equipment or tools. However, most of the young people interviewed commented that they had not been informed that this facility was available to them. Personal advisers confirmed that the subsidy was used on a discretionary basis and that they only informed a limited number of clients likely to secure a 'good job' since they had received (central Government) directives stipulating that expenditure should be contained whenever possible.

The recruitment phase gives a further indication of the process of job matching. A number of employers were not used to recruiting unemployed young people. In fact, many micro-companies had no prior experience of recruitment and reported that the NDYP had given them the opportunity to consider employing a person. The main recruitment channels reported by employers were word of mouth, use of the Job Centre and, more rarely, the placement of an advertisement in a newspaper. Recruitment agencies were only used for recruiting more experienced staff. We found strong evidence that both employers and Employment Service staff sought to minimise the transaction costs involved in the recruitment and placement of New Dealers. In many instances, employers had already registered vacancies with the job centres or were known for employing young staff. These employers were contacted in priority by NDPAs who were under the pressure to achieve placements in a limited period of time.

The majority of employers in Cambridge, Edinburgh and North London reported a low number of candidates per job as opposed to areas such as Birmingham and North Tyneside, where the choice of candidates was much greater. This is not surprising as employers are known to have more applicants in areas of higher unemployment than those situated in tighter labour markets (Brown et al., 2001). In areas with more candidates, employers nonetheless complained that the quality of applicants was poor, which was reflected in negative attitudes during interviews.

In buoyant labour markets, many employers implicitly or explicitly referred to the low employability of New Deal candidates, explaining that they had decided against participating from the outset as they felt that New Deal was not suited to their business needs. Large employers mentioned that they feared that New Dealers lacked appropriate qualifications, and that government officials had not been responsive to their business needs. Some reported having interviewed candidates, but being disappointed with the quality of the applicants so that they preferred to avoid taking risks. Of course, when labour markets get tighter, some employers start experiencing recruitment difficulties due to labour shortages. They are thus more willing to alter their usual human resource practices and consider a broader pool of candidates. As a result, the NDYP has the potential to act as an important element of a learning process that may change traditional practices:

> There must be something for us to take from that because what we are seeing are people who would normally fail, but given a bit of a hand they're actually passing later on and they are now doing a good job for us. Maybe there's something for us as an employer to think about. Are we getting our recruitment right? (employer, Edinburgh)

## A Pool of 'Cheap Labour'?

Another dimension of the disparities within the New Deal employment option concerns the earnings received by young people. Early concerns were voiced that workfare schemes such as the NDYP would offer employers a means to subsidise their wage bill (Peck, 1999). As one employer we interviewed argued, 'there's a danger that employers will exploit the system because they're getting cheap labour' (SME, Cambridge).

The National Survey of participants found, quite unsurprisingly, that those in subsidised jobs tended to earn less than those in unsubsidised jobs (Bryson, Knight and White, 2000, p. 91). These findings are based on comparisons according to occupational status in different establishments and do not provide comparisons of practices within establishments having employed both categories of recruits. Although our survey concentrates on companies who have employed young people in the subsidised option, some employers, especially in the public sector stated that they had recruited under both categories for similar positions at the same starting wage. However, progression up the pay ladder was quicker for those in unsubsidised employment because they held prior qualifications that enabled them to compete in internal recruitment competitions. Arguably, this supports the idea of a 'low skill – low pay' trap, yet in the great majority of cases we found that employers who had offered permanent employment subsequent to the six months New Deal period had also increased the pay offered. Virtually all employers who had retained their New Deal employee(s) or who were planning to do so reported an increase in pay rates of about 10 per cent. Hence, since tenure appears to contribute to the probability of obtaining higher wages, it is important to understand the cases where New Dealers stayed in employment with the same employer after six months, and we return to this later. National Survey findings show that in the early stage of New Deal, 62 per cent of recruits had been retained after the subsidy ended at six months (Hales et al., 2000). However, there are indications that this retention rate fell as the programme matured.

Tables 5.3 and 5.4 show wages paid *during* the six months subsidised employment period. Our data on wages below £3.00 per hour is below that of the national survey that was carried out one year earlier, and before the effects of the introduction of the national minimum wage in April 1999.[2] This probably explains the difference between the 21 per cent of New Dealers receiving less than £3 per hour reported in the national survey of employers, and the average of 8.5 per cent for our five case-study UoDs.

Birmingham and North Tyneside displayed a greater proportion of New Dealers with low rates of pay. Birmingham had the highest reported hourly wage, paid to trainee engineers by an international car manufacturer, but this was a noticeable exception in this area. There were also examples of very low rates of pay averaging £1.50 per hour in North Tyneside. In several instances, in contravention of New Deal rules, employers considered that they did not have to pay a wage and did not supplement the basic level of subsidy that was paid to young people as a training allowance. In other instances, lower earnings were topped up by non-wage benefits such as free accommodation, as reported in Cambridge. The relationship with housing proved particularly important in London, where the gap between high housing costs and low earnings acted as a disincentive for young people to consider paid work. In our London UoD, it became customary for NDPAs to encourage New Dealers to take-up part-time employment, and so-called 'passported' benefits were calculated so that work could be combined with a continued eligibility to housing benefits.

Compared to their usual recruitment of young people, employers did not appear to alter the usual terms of pay offered to this particular age group,

**Table 5.3** Rates of pay and weekly hours of work in subsidised New Deal 18–24 jobs, by local Units of Delivery

| | Birmingham | | | Cambridge | | | Camden/ North Islington | | | Edinburgh | | | North Tyneside | | |
| | W | H | HR | W | H | HR | W | H | HR | W | H | HR | W | H | HR |
|---|---|---|---|---|---|---|---|---|---|---|---|---|---|---|---|
| Median | 150 | 37.5 | 4 | 158 | 40 | 4.22 | 127 | 35 | 4.25 | 144 | 37 | 3.80 | 122 | 37 | 3.45 |
| Highest | 350 | 41 | 9.33 | 200 | 40 | 5 | 250 | 40 | 7.14 | 192 | 40 | 5.20 | 173 | 52 | 4.83 |
| Lowest | 60 | 34 | 1.76 | 115 | 35 | 2.87 | 64 | 16 | 3 | 100 | 30 | 2.70 | 60 | 30 | 1.5 |

Note: W = weekly earnings (in £), H = hours worked per week, HR = hourly rate (in £).

**Table 5.4** Percentage of New Deal 18–24 recruits receiving different hourly pay rates

| Hourly wage | Birmingham | Cambridge | Camden/North Islington[a] | Edinburgh | North Tyneside | Average (5 UoDs) | National Survey of Employers[b] |
|---|---|---|---|---|---|---|---|
| Less than £3.00 | 12 | 6.5 | 0 | 6 | 18 | 8.5 | 21 |
| £3.00–£3.99 | 35 | 26.5 | 31.25 | 53 | 54.5 | 40 | 50 |
| £4.00–£4.99 | 47 | 53.5 | 31.25 | 29.5 | 27.5 | 37.5 | 21 |
| Over £5.00 | 6 | 13.5 | 37.5 | 11.5 | 0 | 14 | 8 |

Notes: [a]Results for Camden / North Islington to take with caution because of lower weekly hours, see table 4.3.
[b]Data in the last column are extracted from the National Survey of Employers, Hales et al. (2000, p. 63).

unless vacancies were classed as apprenticeships from the onset. This is because New Dealers were typically recruited into entry-level jobs. It is therefore not surprising that our findings concerning the wage rates offered to New Dealers are similar to those obtained via a national survey of young participants, which shows that earnings were in the bottom decile (Bonjour et al., 2001, p. 40). Furthermore, these outcomes are consistent with a study of recruitment and retention in lower-paying labour markets, which shows that wages rates in those types of labour markets are in the bottom fifth of the overall distribution of earnings and that there is a trade-off between pay and turnover (low-pay results in higher turnover, and vice versa) (Brown et al., 2001).

When employment contracts were broken before the end of the subsidised period, in the majority of the cases this was a decision taken by the employee. Young people were often reported to have vanished from the workplace without prior notice or warning. The lack of attractiveness of both the earnings and the nature of the jobs could in part explain why young people had not developed the sense of loyalty and responsibility that contractual relationships are expected to trigger.

## The Role of the Job Subsidy

Wage cost subsidies have been one of the key elements of active labour market policy and they represent one of the most direct methods of increasing unemployed people's chances of competing in the labour market. The New Deal wage subsidy of £60 per week is intended to help facilitate the entry and integration of young people in the workplace by reducing the initial costs of recruiting. The subsidy can be regarded as a form of *compensation* to offset an initially lower level of productivity and the costs of helping the young person adjust to a working environment. The subsidy may also be seen as an *insurance* against the risk that the person recruited would turn out to be unsatisfactory in some ways (Hasluck, 1999b; Hales et al., 2000, p. 141).

These two rationales for offering a financial incentive are indeed recurrent in the marketing of New Deal by the Employment Service. There have been some variations in how the subsidy was presented to and received by employers according to the type of organisation, and the type of local labour market concerned. In buoyant areas, the Employment Service appears to have put an emphasis on the *contractual* dimension of New Deal: the subsidy has tended to be presented as a compensation for the additional effort expected from employers, with the view that the New Deal placement would then adjust to the workplace and be retained. In more depressed labour markets, where the Employment Service is under more pressure to place young people, the subsidy was more likely to be presented as an insurance, not solely against the risk of the person proving unsuitable, as Hasluck and others suggest, but also against the risk of the company not achieving its planned expansion and being as a result unable to keep on the person, *despite* their proven and potential workplace related abilities. While UoD managers later came to realise some of the mistakes and readjusted their marketing strategies, below are two contrasted examples of how the subsidy was sold to employers:

> The subsidy was presented to us as something to pay us back for the extra effort that is involved in helping somebody without the qualification (employer, Cambridge).
>
> They were sort of saying, 'Take a risk: it's only six months and you don't have to have her' (employer, Birmingham).

Unsurprisingly, many employers expressed the view that the subsidy was a 'carrot', a 'sweetener' or a 'nice little perk' that enticed them to recruit young unemployed people and diminished the risk of recruiting. In many cases the payment of the subsidy appeared to prolong a 'wait and see attitude' among employers who were unsure about whether to retain the employee.

There is, therefore, little doubt that the wage subsidy had a significant impact on employment practices. It is clearly important to assess the positive net effect of the subsidy on employment, and consider whether the outcomes would have occurred in the absence of the programme. In other words, a key question is whether the job offered to a New Dealer is 'additional' and whether the intervention in favour of this target group may be detrimental to other jobseekers (Hasluck, 1999b; Hales et al., 2000, p. 141).

The literature on active labour market policy points to various effects that may arise from wage cost subsidies (Disney and Carruth, 1992; Calmfors, 1994; Hasluck, 1999b). The three most significant effects are deadweight, substitution and displacement. Deadweight denotes that the economic activity or jobs would have occurred anyway, even if the programme had not been in place. Substitution effects relate to the proportion of New Deal participants who were recruited to subsidised jobs in the place of other individuals seeking work or employed with the same employer. Displacement describes the way in which the programme may produce employment growth in participating firms at the expense of employment in non-participating firms.

Estimating what would have existed in the absence of the programme is complicated and considerable caution is required. Some idea can be gained by an assessment of the situation prior to and after recruitment, and by looking at employers' perceptions of a possible counterfactual. However, this method of collecting information requires employers to make retrospective and hypothetical judgements, which some may find difficult to formulate, while others may seek to conceal their real motivations. Some employers revealed that they had participated in the New Deal to employ a person they knew and to pass the benefits of the subsidy on to them. While additionality is regarded as a positive effect, since it corresponds to net job creation, it should be recognised that if employment is not sustained, it may be a reflection of rent-seeking behaviour, which is why some employers may have been reluctant to reveal their motivations.

Early evidence from qualitative research has shown that, in general, employers had no intention to use the New Deal as a means of subsidising their wage bill (Snape, 1998). This suggests that employers were not primarily using New Deal to recruit staff who they could afford to employ in the first place, implying that the level of deadweight was low. However, this type of 'soft' evidence remains highly controversial, as one can assume that when questioned, employers would be reluctant to admit to a rent-seeking strategy. Our interviews found several examples of establishments who retained New Dealers for a period of between five and seven months, and were then unwilling or unable to offer them a longer contract. There tended to be three types of justification for the termination of a contract at this stage. Some employers admitted that the termination of the contract was due to their own incapacity to sustain the position despite genuine hope for business expansion. Others argued that the young person had committed several mistakes or lacked the required discipline and that such behavioural problems had only become visible at a late stage, which explains why they had decided to keep them in post throughout the subsidised employment period. A third category of employers reported that it was the young person's decision to leave.

The National Survey of employers found that 69 per cent of the subsidised jobs in the NDYP was deadweight (Hales et al., 2000). Moreover, deadweight was found to be lower in the voluntary sector than in the private and public sectors, and also slightly lower among firms operating in a local/regional market for final products/services as opposed to those operating in a national or international market. Finally, deadweight also appeared to be related to the size of employers. It was lower in micro-firms (employing fewer than five individuals) than in SMEs and large companies. Moreover, in our sample, employers in the early stage of starting their business or highly dependent upon seasonal business fluctuations such as in the building and leisure industries – many examples of which were found in North Tyneside and North London – proved to be very responsive to the subsidy.

It was not surprising that deadweight was high, since earlier evaluation of the Workstart Pilots, which had many similarities to New Deal, consistently found deadweight in excess of 70 per cent (see Hamblin, 1997). The most common way of measuring deadweight is to ask the employer whether the vacancy would have existed in the absence of New Deal. However, this is fraught with difficulties, as some employers reported that the vacancy would have existed but under different conditions, such as part-time employment or at a reduced rate of pay. We would class this type of response under 'partial additionality' (our interpretation of partial additionality differs from Hales et al., 2000, who class under partial additionality a position that was brought forward).

Respondents who reported that the vacancy would have existed under the same conditions could be further divided between those who are certain that they would have employed the same person, those who were certain that they would have employed someone else (such as an older worker), and those who were unsure. Our definition of substitution included only those employment contracts that would have definitely been offered to other candidates, which means that actual substitution may be higher.

Table 5.5 shows that deadweight was high; the most common trend was to turn part-time jobs into full-time jobs, especially among small employers. There was evidence that some employers delayed recruitment to take advantage of the subsidy when wanting to employ a specific person that they knew already who had not been unemployed for the qualifying period of 6 months. At the same time, there was also evidence of employers who signed up to New Deal but did not recruit immediately. This suggests that employers were not rushing recruitment to take advantage of the subsidy if they have no need to employ someone or if they were keen to wait for the 'right' person.

One of the assumptions made regarding local labour markets was that substitution would be higher in depressed areas (Holtham et al., 1998). Overall substitution has been quite negligible at 14 per cent of the cases (Hales et al., 2000, p. 147). Most employers reported that they approached the NDYP because they were looking for young candidates that they could train at a relatively low cost and that the vacancies would have been unsuitable for an older and more experienced worker.

## Training Provision under the NDYP

The distinctiveness of the NDYP, and of the subsidised job option in particular is reflected in the emphasis on the training element. The rationale for training is obvious: job seekers are at a relative disadvantage in the jobs market if they do not possess the skills that employers require. No qualifications and poor or low-level qualifications are associated with an increased risk of long-term unemployment (Hasluck, 1999a, p. 41). Of course, training may benefit employees in terms of access to employment,

**Table 5.5** Estimates of additionality and deadweight effects in New Deal subsidised placements (percentage of vacancies)

| | Birmingham | Cambridge | Camden/North Islington | Edinburgh | North Tyneside | All UoDs |
|---|---|---|---|---|---|---|
| Additionality[a] | 23.5 | 38.5 | 18.5 | 47 | 65 | 38.5 |
| Deadweight[b] | 23.5 | 31 | 44 | 35 | 15 | 29.5 |
| Partial additionality[c] | 6 | 23 | 25 | 6 | 5 | 13 |
| Substitution[d] | 47 | 7.5 | 12.5 | 12 | 15 | 19 |

Notes:
[a] There would be no vacancy.
[b] There would be a vacancy, probably filled by the same candidate.
[c] There would be a vacancy, but under other conditions (e.g. part-time or lower remuneration).
[d] There would be a vacancy, but filled by a different candidate (e.g. with more experience).
Source: Authors' Survey.

and also increase their potential to earn higher wages (Green, 1999). Training also has associated benefits for the local economy. Not only is unemployment reduced in the short term, but in the longer term, skills shortages and gaps will be reduced. This is particularly important for companies operating in depressed areas that need to improve their relative competitiveness, output, employment and profitability levels. However, the potential benefits do not offer a guarantee for optimal outcomes: the structuring of training can lead to the reinforcement of labour market disadvantage. Access to the 'best' training places can be restricted to applicants who employers regard as a good risk, so that differential access to jobs may be a reflection of differential access to training places (Hasluck, 1999a, p. x).

Employers who signed up for the New Deal were made aware that training provision is compulsory, and in addition to the wage subsidy they are offered a £750 allowance to cover the costs of training. In principle, the delivery of training is validated by a formal administrative procedure that consists of agreeing on an individual training plan. But actual practices have differed hugely across UoDs. In some areas, the initial visits of local Employment Service staff during which employers sign their local agreement have been regarded as sufficient, although in the majority of instances, training preferences and provision were still unclear at that stage. In North Tyneside, employers were relieved from the responsibility of arranging a training course. The New Deal management officials tended to offer an in-house package right from the onset. While this type of integrated package was often valued by small employers, it could be suggested that this was also to the benefit of the New Deal team. The leading agency for NDYP was an IT training provider that 'offered' to act as the course provider, which meant that they could reap the benefits of a 'quasi-monopoly' on training provision. Edinburgh appears to have been the most stringent in that no subsidy was paid until the individual training plan that the employer was left to fill out with his/her employee was returned.[3]

Past managerial practices in relation to training appeared to have had a major bearing on the arrangements made for New Deal recruits. There is overwhelming evidence from the literature that employers in Britain have tended to disengage from training as they have externalised risks on to individuals (Grimshaw et al., 2001). This does not mean that employer-sponsored training does not exist, but it tends to be short term: less than three months (Green, 1999). This tradition is perceived as central both by the agents delivering the programme and by employers themselves:

> Certainly the training element is very important because we've traditionally had this mismatch between what employers want and what employees have to offer. To be honest, it has been hard to get employers to accept that one of the

things that they need to do in order to get the people they want is to be prepared to invest in training them (ES regional manager, 3 December 1999).

We all know that in the last fifteen years there has been a problem with companies financing training. It's survival of the fittest at the moment and unfortunately training is one of the last things on the agenda and that is a pity (SME, Birmingham, 27 April 2000).

One way of gauging the quality of training is to investigate whether it leads to any form of qualification; but this in itself is problematic given the lack of transparency in the National Vocational Qualification (NVQ) system and the criticism that surrounds it (Green, 1999). It is certainly the case that many of the employers and employees we interviewed saw Scottish Vocational Qualifications and NVQs as irrelevant to their needs and described them as a 'waste of time'.

A common argument is that the reluctance to undertake training is linked to market failure, which in the UK has created a 'low-skills equilibrium' (Finegold and Soskice, 1988; Hasluck, 1999a, p. 72). This arises from employers' perceived inability to capture a return on their investment (the fear of labour poaching), as well as wage rigidities that prevent the unemployed paying for their own training (for instance by working as trainees or apprentices at low wages) and capital market imperfections, which constrain investment and innovation and thus the demand for skills. One employer argued that

I'm getting sick to death about hearing people say, 'People are my best asset.' They are my best asset, but they're also my worst enemy because I invest a lot of time in them and money to train them through and then they're poached very rapidly and easily. When you are a large company you can provide career structures, safety, security, although security doesn't exist these days. We cannot provide them the glorified career path in that particular way (SME manager, Birmingham).

According to ES officials, the compulsory training element was a means to sift employers; those who intended to participate in New Deal merely to take advantage of the subsidy withdrew their interest when it was explained that there was an obligation to provide a day's training per week. Furthermore, official New Deal literature gives the impression of relatively homogenous and effective training practices and outcomes. However, despite the fact that three quarters of New Deal employees in the subsidised employment option were reported to have been enrolled on a training course (Bryson, Knight and White, 2000), we found varying interpretations among employers as to what constitutes training and the degree of commitment that should be displayed. Unsurprisingly such attitudes often mirror the occupational categories and the sector of activity in which young people were employed. Managers in the hospitality sector (in all areas, but in North London particularly), for example, had conspicuously dismissive and negative views of training requirements, as reflected below:

> We need some porters. We will teach them. For a couple of weeks we will fully train them (employer, North London).
>
> We don't have the time to sit here and train people. Kitchen portering, you can do it on the job. We can tell them what to do and what not to do (employer, North London).

The responsibility for making training arrangements was often left with the employer, resulting in a wide range of attitudes and a discrepancy in training outcomes. Large employers tended to rely on their well-developed in-house programmes, although some took a restrictive view of training and limited its interpretation to an induction course on health and safety issues. Arranging training proved particularly problematic, for small employers and their attitudes differed significantly. Some small employers felt that it was young people's responsibility to source their own training and complained that their employees had not been proactive. In contrast, those employers with well-defined and specific needs often complained that the type of training required was not available in the official providers' lists proposed by the ES. It was often the case that specific training with accredited boards had to be sourced outside the local labour market. This was for example the case for a film production company located in North Tyneside for whom a course with the BBC would prove the most appropriate type of training for his employee. Yet the employer argued that the delivery agents did not recognise this training as eligible. In all areas, on-site training was predominantly used, since employers proved reluctant to give the young recruits time off. The most common practice consisted of completing a portfolio towards an NVQ with three to four visits by a tutor over the six-month period.

One innovative approach, which was only used in Edinburgh, was the use of the training subsidy towards the cost of driving lessons. While this practice was originally not allowed under New Deal rules, it later proved relatively popular among both employers and young people and was subsequently widely adopted in other UoDs.

## A Typology of Employer Participation

Cultural interpretations provide useful insights into the character of enterprise strategies (Schoenberger, 2000). Employers approached the NDYP with varying objectives and dispositions, and experiences of the programme have been uneven. A typology of employer participation can be constructed according to what might be termed an 'organisational mode'.

Table 5.6 outlines a typology of employer engagement in New Deal according to three distinctive ideal-types, based on the organisational character and priorities of employers and their response to either market, hierarchy or social signals. The idea of a relational type draws on the concepts of 'relational assets' (Storper, 1996) and 'social responsibility', which highlight the intangible non-market resources in economic life. Such concepts have featured prominently in non-profit sector studies (see OECD, 2003), but of course they are also important to many other types of employment. When asked about this notion, three types of responses were given by interviewed employers: some expressed a strong degree of social conscience and were used to participating in a range of other local community initiatives, others limited and internalised social responsibility to their own staff and often to one particular individual, while a third category proved highly dismissive of this concept and preferred to emphasise their responsibility to make profits and pay taxes.

In the relational mode, the decision to participate in New Deal tended to be based upon individual relationships and social ties, as exemplified by the case of family businesses. In both buoyant and depressed labour markets, we found evidence of employers who employed their own children or friends through the NDYP. In some cases, this strategy was suggested to employers by ES officials. In other cases, 'enlightened' managers with a relatively strong degree of autonomy decided to 'do their bit' for the community. Therefore, an informal mode of recruitment was chosen, since the young person was relatively close to, or known to the employer.

This was the case with a shop-owner selling surfing equipment who employed a friend through New Deal. This employer argued:

> We just worked it between us and it worked nicely. We had a little bit of fun as well. I said I go for a surf for an hour and then you can go for a surf for an hour. We had quite a good deal (employer, North Tyneside).

The outcomes for those young people experiencing this relational mode tended to be higher job satisfaction and a sense of ease in the workplace. A young person in such a situation reported:

> What is good is that (employer's name) gives you that space to kind of feel your way around to see what you feel that you can do and what you feel that you need to improve on. So within time, from then to now, I've developed a lot and been given more responsibility (young person, aged 21, London).

**Table 5.6** A typology of employer participation in the NDYP and its outcomes for young people

| Organisational mode | Prevailing Rules and Conventions | Co-ordinating Mechanisms/Incentives | Type of Recruitment | Outcomes for Young People |
|---|---|---|---|---|
| Opportunistic | Belief in market exchange principles, rent-seeking, risk adversity | Price (hence high sensitivity to the wage subsidy) | Formal/competitive | Young people used as commodified labour; high risk of churning |
| Authoritative/ rational | Application of vertical directives, relative lack of autonomy in decision-making | Quotas and targets for New Deal recruits, hierarchy and subordination | Formal/rule-based | Investment in the most employable and job-ready young people; Sustained employment more likely |
| Relational | Mutuality, reciprocity, trust, sense of duty and responsibility | Individual and social ties | Often informal | Investment in young people's skills and future employment but not always sustainable |

Economic sociologists have stressed that economic behaviour is embedded in networks of interpersonal relations, and is therefore crucially influenced by aspects such as reciprocity, trust and co-operation (Granovetter and Swedberg, 1995). Such intangible factors appeared to be influential among participating companies. Reciprocity, for example, was a recurrent theme. As a principle, reciprocity reflects an established relationship between a group of individuals, following a series of mutual gifts and favours (Gouldner, 1960; Mauss, 1990). The interviews clearly demonstrate that such norms were taken into consideration:

> I'm keen to help youngsters to find the right niche in life because I know how hard it is for people to get the right niche. I think *it's good for both sides* (employer, Cambridge; emphasis added).
>
> We should invest in local people because if they are suitably upskilled and suitably supported and committed to, they will stay within the organisation. *So it's a two-way investment* (public sector employer, Birmingham; emphasis added).

In some contrast, the authoritative/rational type was mainly apparent in large public and private sector enterprises – such as manufacturing companies, hospitals, transport and the police. In this mode, a set number of vacancies open to recruitment tended to be ring-fenced for New Deal recruits in order to comply with agreed targets. Such employers were most likely to sit on New Deal Strategic Partnerships and had generally introduced well-structured mentoring programmes at establishment level, pairing a New Deal recruit with an experienced member of staff. These employers reflected that NDYP represented a form of long-term commitment and investment. They expressed an obligation to 'put something back' into the local community, although this was also of course part of the corporation's public relations strategy.

In depressed labour markets, New Deal strategic partnerships made a particular effort to engage large public sector organisations (which could be classed as belonging to this authoritative type). A study of public sector employers argues that this type of employer can 'set the example' for other employers (Nathan, Simmonds and Ward, 1998), as reflected by a Birmingham-based employer:

> The reason why things like the New Deal work is probably because large local companies have got that ethos that we will work with the Employment Service to actually deliver and deliver well, and therefore we can probably lead the way for some of the smaller companies to do it (large employer, Birmingham).

However, in tight labour markets where the numbers of New Dealers was much lower, the participation of public sector employers was limited, and the recruitment of young people in this sector was lower. In Cambridge, a manager working for a large public sector organisation reported to having recruited only one young person and felt that this represented a failure:

> I don't think I'm the most committed to training and development for youngsters in this area and it may be it is a coincidence that I'm the one manager who's gone ahead with this. But I think that in the health service generally the welfare of employees is right up there at the top of the list more than a lot of employers and that in itself would suggest that that shouldn't be a reason for the failing on this sort of scheme (employer, Cambridge).

This example highlights the lack of co-ordination in some large bureaucratic organisations, where line managers were often not aware of the possibilities of using New Deal as a mechanism for recruiting.

Employers in the non-profit sector typically attached greater priority to the interests of the young people. However, the participation of these 'community businesses' in the employment option was low, since these potential employers were essentially mobilised for participation in the Environmental Task Force option, especially Intermediate Labour Markets (ILMs), which are often seen as suitable for temporary one- or two-year placements. However, our research indicates that many community businesses, which do not consider themselves to be proper ILMs, also have the potential to offer rewarding employment at reasonable wage rates. The focus on private sector employment means that the New Deal has not been effectively marketed to them. Examples in Europe show that this sector can make an important contribution to the labour market integration of young people because the quality of the jobs, and the mentoring and development young people obtain is often superior and valued by employees (Laville, 1996).

In the 'opportunistic' mode (which characterises the great majority of employers), the subsidy clearly tipped recruitment in the favour of the NDYP. It has been argued that many SMEs assume that New Deal recruits will be 'job-ready', implying that their supervisory and training capacity is likely to be below that offered by large employers (Tavistock Institute, 1999, p. 41). We found varying levels of commitment and provision for the training and personal development of young people. In the opportunistic mode, employers had rarely developed a specific strategy concerning the development of their New Deal recruits. They were often responding to short-term cyclical pressures. Fluctuations in business activity and profitability were the determining criteria and these resulted in the recruitment of a 'contingent' workforce (to use Peck and Theodore's [2000a] term) with lower levels of security or career progression compared to those employed under the relational or authoritative mode.

Moreover, the issue of 'absentee employers' proved to be relatively significant. It appears that small retail outlets that operate within the opportunistic mode displayed a poor sense of supervisory responsibility, more so in North London and North Tyneside than in other areas. For many shop owners, the presence of young people on the premises was a means to gain some time off for other activities. One employer in North Tyneside reported that he had opened his shop as a means of supplementing his own income from his salaried activity, which involved shift working. So while establishing this new outlet, the employer continued his main employment. The New Deal employee was expected to be present in the shop in the morning until the employer came back from his shift in mid-afternoon. Extreme cases were reported where young people were left unattended on the premises and where employers would merely attend to collect the day's receipts:

> I was left by myself up until 7 pm from 9 am. It was like 'I'm just popping out' but then you'd never see him and he used to ring at quarter to seven asking, 'Is everything all right?' Basically, he just couldn't be bothered (New Dealer, North Tyneside).

While estimating the local density of the various types of firms was not possible, firms displaying each of the three organisational characteristics can be found in all types of labour markets. The crucial point, however, is that the prevalence of 'opportunistic' employers relative to the number of New Dealers clients tended to be higher in depressed labour markets.

## Conclusions

Our results suggest that two interdependent dimensions of variation have been evident in employers' involvement in the programme. First, there are major variations across the five case-study areas, both in terms of the employment opportunities available to young people and the wages offered. Tight labour markets proved to be more 'human capital' oriented with higher levels of employer commitment to the development and training of New Deal recruits. Second, however, there were also variations between the different types of employers involved within different labour markets, and we have tried to summarise these through a simple typology of motivations and practices. Thus when the participation of large public sector employers and non-profit organisations was secured in more depressed labour markets, this could produce a beneficial and progressive, if ultimately limited effect. But while employers have called for a greater screening and sifting of candidates by the new Deal agencies, the reverse logic, which would have involved the sifting of employers allowed to receive the subsidy, was not applied. Strategic partnerships and New Deal policy-makers did not consider stronger employer screening and diagnosis prior to referring young people to interviews.

Participating employers did not enter the New Deal on a level playing field. Many small companies, in particular, had either no or very little experience of recruiting staff, particularly long-term unemployed youth, and were thus ill informed and ill prepared for the level of commitment demanded by such participation. For many small employers, recruitment and training are far from being conceived as strategic, as they are preoccupied with surviving in the face of strong and immediate business competition. We have seen that where employers had more strategic orientations, greater efforts were made to integrate youngsters in the firm. In this sense, the programme has offered employers a significant opportunity to engage in a learning process, particularly concerning co-operation and liaison with external actors such as the Department for Work and Pensions, but also colleges and training providers.

In addition, local disparities have been apparent in the effects of the wage subsidy. Unsurprisingly, the subsidy has produced more deadweight and substitution in tighter labour markets such as Cambridge, thus tending to redistribute jobs. While additionality has been greater in slack labour markets, this has not necessarily been reflected in high-quality jobs and training with subsequent long-term career prospects. On the contrary, the majority of opportunities – with some notable and rare exceptions – have been manual unskilled and semi-skilled jobs. So while there has been an element of job creation in those areas, the quality of the placements and wages offered to young people has, to a great extent, led to a reproduction of past practices and has not fundamentally altered the dynamics of the local labour market.

In hindsight, a different design for the level and payment of the subsidy might have produced greater impact over the longer term. For example, the payment could have been made over a longer period of two or three years, with a declining rate being introduced after the initial critical 'settling in' period. Furthermore, the type of financial incentives offered could have been aligned to local labour market conditions. This could have involved replacing the wage subsidy by a housing subsidy or mobility grant in tighter labour markets, since such labour markets are already experiencing labour shortages. It is clearly rather anomalous to offer employers an incentive to recruit when they are already experiencing difficulties in finding staff through their usual recruitment channels. Moreover, several employers mentioned that the difficulty resided principally in their ability to recruit suitable staff. Attitudinal characteristics and basic skills were seen as key determinants of employability, rather than qualifications, experience and technical competencies, which has major implications for the way in which the educational system should prepare young people for their school-to-work transition.

In sum, this chapter has demonstrated that employers have been encouraged by a complex combination of economic, social and political incentives to engage in the New Deal for Young People workfare programme. However, in the main, they have stuck to their established labour practices and organisational priorities, so that the programme has tended not to produce significant improvements in the prevailing character of recruitment, training, staff use and development. We have argued that it is necessary to disaggregate the 'big picture' and uncover the multiple realities behind New Deal employment contracts. In most cases, however, the experience of New Deal often proved to be one of unmet expectations among both employers and young people, or to put it differently, between the aspirations of youth, and the structure and operation of local labour markets.

# Chapter Six

# Localising Welfare-to-Work?

## Local Flexibility and the New Deal for Young People

From its inception the New Deal for Young People has been dressed in a language of localism and decentralisation. It has been presented as being at the forefront of the Government's intention of marrying national mainstream labour market policy with local flexibility (Millar, 2000a). The Government has been at pains to emphasise that the NDYP involves an important degree of decentralisation and that it is delivered by local partnerships between the Job Centre Plus (formerly Employment Service [ES]) and other agencies, which 'plan the delivery of the New Deal locally making best use of their knowledge, understanding and expertise in local labour market matters' (DfEE, 1998, p. 2). At the time of its launch, the chief executive of the Employment Service, for instance, stated that there should be 'the maximum degree of flexibility' for the partners to decide how they would plan and contract the delivery of the New Deal programme (Education and Employment Committee, 1998, p. 19; see also Brown, 2001, p. 3). More recently the Government's plans to reform and build on the New Deal once again promise to establish more local flexibility in order to tailor solutions to the needs of individuals, employers and local areas. Within a national framework of standards, both New Deal managers and personal advisers are to be given greater local discretion and devolution (DWP, 2004a).

The rhetoric suggests that we have witnessed a paradigm shift from welfarism to workfarism, involving a rescaling or *localisation* and *decentralisation* of labour market policies and agencies (see OECD, 1999). According to Finn (2000, p. 1):

> the new approaches involve radical changes in traditional welfare and employment agency bureaucracies. In most countries this has been coupled with

decentralisation and the increased use of local partnerships and organisations in designing and implementing new 'welfare-to-work' programmes.

Such claims are usually supported by the belief that the spread of 'welfare-to-work' policies represents the diffusion and imitation of US workfare policies, which are characterised by a high degree of agency decentralisation, local experimentation, as well as local differentiation in policy outcomes. According to Theodore and Peck (1999, p. 489), for example:

> In order to facilitate the diffusion of new approaches to welfare reform, governments are engaging in the decentralisation and localization of delivery systems, adopting new roles themselves as orchestrators and animateurs of local experimentation. This is certainly the case in Britain and the United States, where there are strong similarities in the conventions, practices and discourses of welfare-to-work programming under what remain quite distinctive forms of decentralisation/defederalization.

It is undeniable that there is widespread and growing enthusiasm for more decentralised labour market and welfare policies. Active labour market policy tends to involve an emphasis on the benefits of decentralising labour market interventions and establishing a stronger local dimension to their design and delivery (OECD, 1998, 1999). It is argued that a local dimension allows labour market policies to respond more closely to the variability and diversity in local labour market conditions (Campbell, 2000; Green and Owen, 1998; Finn, 2000). In addition, this enthusiasm for 'local flexibility' has been associated with a critique of existing welfare systems, which contends that centralised state bureaucracies have been monolithic, excessively top-down and inflexible. In this account, welfare policies are criticised for having failed to respond adequately to local and individual needs as a result of 'departmentalism', vertical structures, and uncoordinated or 'silo mentalities' that lack cross-cutting links (Social Exclusion Unit, 1998). Finn (2000) argues that such top-down fault-lines produce a local institutional incoherence in which different agencies with different funding streams pursue different aims.

However, we should be wary about exaggerating the degree of decentralisation implemented in the New Deal. A creeping trend towards 'central government localism' in the delivery of active labour market policies began in the 1980s under the Thatcher governments, with the development of the so-called 'trainingfare state', as represented by the New Training Initiative, Youth Training Scheme and especially the local Training and Enterprise Councils (see Jones, 2000). All of these involved a largely rhetorical shift away from a centralised monolithic policy model towards a greater element of local flexibility of response and implementation – within centrally controlled guidelines and governance structures. Does the New Deal represent more of the same? Peck (1999), for example, noted that despite extensive local consultation, most of the key design parameters were nationally determined and local flexibilities were conceded only at the level of programme delivery, not design. Bennett and Payne (2000) went further and argued that the strategy and implementation of New Deal have largely followed a process of imposing a national template. According to Trickey and Walker also (2001, p. 203), 'the government has . . . stressed the scope for local variation and discretion although the 'top-down' system of delivery and control of social security in the UK has *not* been substantially challenged by this initiative' (their emphasis). In international comparative terms, social assistance in the UK has been highly centralised (see Gough et al., 1997), and this policy inheritance has shaped the NDYP. Lødemel (2001, p. 310) noted that the programme is much more centralised than workfare programmes in the US, Germany and Norway. However, he added that 'While New Deal is the most centralised programme of the seven countries considered here, it is more subject to local variation than the highly centralised social assistance found in the UK in the early 1990s.'

The aim of this chapter is to examine the degree to which the NDYP has actually been decentralised and to determine how far, and in what ways, it has benefited from local flexibility in programme implementation. In order to do this we use the four key arguments for, and alleged benefits of, labour market policy decentralisation as advanced by the OECD (1999): better *partnership-building and co-ordination* of policies, improved *policy learning and adaptation* to local conditions, greater *innovation and experimentation*, and more *resource targeting* on areas with particular needs (OECD, 1999; 2003; see also Campbell, 2000). The local flexibility and partnership autonomy inherent in the NDYP design was introduced with the aim of realising many of these benefits. We therefore examine the achievements and limitations of the NDYP under each of these headings to consider whether it has succeeded in reaping the potential benefits highlighted by each. The chapter mainly uses interview findings in our five case study areas, with Employment Service officials, local programme delivery partners, option providers, and employers. The interviews sought to establish the degree of local responsiveness exercised by the relevant partnerships, and to highlight some of the key obstacles to further local sensitivity and adaptation.

Our findings suggest that while the partnership-approach and contractualism of the New Deal have allowed a certain degree of local discretion, there have also been strong bureaucratic and financial limits on the degree to which local flexibilities have been introduced and local needs met. The Job Centre Plus hierarchy has kept a firm grip on the programme's implementation in order to ensure that its standard design and funding principles have remained intact. The chapter argues that decentralisation has not been straightforward, as it has been one element of a wider modernization of welfare delivery, based on the introduction of managerialist practices and new public management principles (Clarke, Gewirtz and McLaughlin, 2000; Cutler and Waine, 2000; Cochrane, 2004). Crucially this modernization has contradictory implications for local delivery and flexibility. On the one hand, it has emphasised the importance of cultural change in service delivery and the importance of innovative local leadership, contracting out and partnership with other local agencies, and learning from local innovations. But on the other hand, this has been framed and regulated by the implementation of other key new managerialist techniques, such as performance measurement and evaluation, expenditure controls and output targeting, that have circumscribed the decentralisation of authority and initiative. Decentralisation has been limited by the top-down manner in which the DWP (formerly DfEE) has imposed its supply-side strategy and reform of the culture of New Deal delivery. This 'modernisation' has been a centrally driven reform programme in which accountability has remained vertical and in which responsibility continues to flow mainly along the lines of a hierarchical bureaucracy. In such a reform process, allowing local agents to deviate by setting their own agendas and deviating from prescribed targets and priorities was seen as far too risky. However, the chapter then considers whether the greater local flexibility signalled in recent reforms to the New Deal will resolve these contradictions and will significantly increase its decentralisation, yielding the predicted benefits.

## Partnership-Building and Co-ordination

There are several possible and often interrelated ways of decentralising active labour market policy. One is to grant local and regional units of the relevant agencies more financial and operational autonomy, while another is to create local partnerships with responsibility for formulating relevant plans of action and for directing their implementation. The design of the NDYP did not signal a major decentralisation of power to regional authorities, and the main roles of the ES Regional Offices were to monitor and police the performance of local Units, and to manage contracts (see Table 6.1). Thus there has been no significant element of policy formation at the regional level, as this continued to be subject to national guidelines and directives. At the same time, however, the programme was designed to incorporate and make use of local institutional partnerships. This is unsurprising in the sense that one of the main rationales advanced by supporters of the local delivery of labour market policy is that it can facilitate *partnership-building and the co-ordination of different policies*. The popularity of partnership approaches within 'Third Way' thinking, arose partly from

**Table 6.1** Employment Service managerial responsibilities in NDYP

| ES Territorial Units | Levels of Competencies and Selected Responsibilities |
|---|---|
| | *Overall Policy Design and National Harmonisation* |
| National Head office based in Sheffield (led by its Chief Executive and made up of four Directorates) | • Horizontal Co-ordination with Secretary of State for Education and Employment, DfEE ministers, New Deal Task Force, the Benefits Agency, Home Office and further public sector agencies/officials<br>• Planning process: production of 'operational plans': setting of national targets and standards (efficiency, throughput, quality), framework for management information, financial regulations and administrative procedures<br>• Vertical directives and co-ordination: communication of targets, recommendations concerning choice of local partners and of implementation procedures, final decision and approval of successful local bids under the Innovation Fund<br>• Research and evaluation: in-house monitoring and co-ordination with commissioned and independent research institutions and universities.<br>• Large Organisations Unit: Co-ordination with large enterprises |
| | *Regional Policy Delivery and Interface* |
| Regional 9 offices (7 offices in England, Office for Scotland and Office for Wales) | • Directives concerning size and choice of delivery model for individual UoDs<br>• Contract agreement and management: Call for tenders and selection of external agents in PSL models. Processing of payments to training contractors and financial subsidies to employers (job option) selected locally<br>• Horizontal co-ordination with external agencies (e.g. regional government offices, TECs/LECs, etc.) and sharing of good practice with other ES regions<br>• Vertical co-ordination: upwards (feedback to head office, clarification of policy lines) and downwards (supportive function for individual UoDs: information about national guidelines and regulations, organisation of regional ES events and conferences)<br>• Monitoring of activities and performance analysis through Regional Intelligence Units |

*Continued*

**Table 6.1** Continued

| ES Territorial Units | ES Agent Categories: | Levels of Competencies and Selected Responsibilities |
|---|---|---|
| | | *Local Strategy and Policy Implementation* |
| Local (144 Units of Delivery and over 1000 local job centres. Depending on their size, individual UoDs may comprise between 4 to 14 local job centres) | District Manager (DM) | • Formation of Steering Groups and identification of strategic partners<br>• Production of local delivery plans<br>• Contractual negotiations with local training and service providers<br>• Permission and approval of potential transfer of funding between options (discretionary powers) |
| | New Deal Co-ordinator | • Co-ordination between DM and strategic partners, training providers and frontline staff<br>• Co-ordination of innovation and new practices, e.g. joint bids under the Innovation Fund and ESF, intermediary projects |
| | Business Managers | • Reception and sourcing of vacancies<br>• Collection of Local Labour Market Information<br>• Management of NDPAs: their training (towards S/NVQs Level 3 in Guidance), communication about procedures, setting of caseloads |
| | New Deal Personal Advisers (NDPAs) | • Client guidance, referrals to the programme's options, job match against vacancies on LMS, client visits, co-ordination with employers, Gateway and option providers concerning training arrangements<br>• Upward communication: good/bad employer practices, clients referred for sanctions, implementation bottlenecks, etc. |

Note: From 8 June 2001, the ES became part of the newly formed Department for Work and Pensions following the restructuring of the former Department for Education and Employment (DfEE) and Department for Social Security (DSS). For more details, see ES Annual Reports and Accounts: http://www.employmentservice.gov.uk/ English/about_us/annualreport.asp.

the belief that they could deliver 'joined-up' and co-ordinated policies (OECD, 2003; Geddes, 2000; Glendinning et al., 2002). A partnership structure may be able to better co-ordinate local employment policies and pool organisational knowledge and resources. An integrated multi-agency approach is supposed to allow more comprehensive and mutually reinforcing support of individuals (OECD, 1998). It is also argued that partnerships can improve the participation of local community groups and so may be more effective in mobilising people, employers and community groups in support of policy objectives (although for a critical view see Geddes, 2000). A more inclusive approach towards community-based organisations and a stronger sense of local 'ownership' may help secure increased enthusiasm and legitimacy for labour market policies (Campbell, 2000).

In the case of New Deal, different types of partnership have been established, but all depend on an underlying contractualism in which money is given to 'service providers' on the basis of the number of clients referred. This blend of contractualism with partnership-building allows a wide range of agencies to be involved and used but, as we will see, it is not without its internal tensions. Four types of delivery models, based on different forms of contracting were implemented (Tavistock Institute, 1999; Atkinson, 1999) (see Table 6.2 and the definitions given in the appendix to Chapter 3).

When the New Deal was introduced, one of the delivery models was set up in each of the 144 Units of Delivery across the country. Our case studies included the full range of delivery models (the Birmingham UoD was a Joint Venture Partnership, Camden and Edinburgh were Employment Service-led, Cambridge a consortium, and North Tyneside private sector led). While all agencies interviewed reported satisfactory degrees of local consultation prior to the establishment of specific partnership models, the final choice of model was often decided at regional rather than local level (Table 6.1). Nevertheless, it was hoped that partnerships would evolve and mature in an organic fashion. Despite the apparent diversity of the different local 'models', the structure of the programme was basically the same in all areas. Moreover, management effectiveness and style varied significantly within each type of model and this variation has been widely judged as

**Table 6.2** Numbers of Units of Delivery by type of delivery model

| Type of Delivery Model | Number of UoDs | % of Total |
|---|---|---|
| Employment Service Individual Contracts | 108 | 75 |
| Joint Venture Partnership | 13 | 9 |
| Consortium | 13 | 9 |
| Private Sector Led | 10 | 7 |

having more of an effect on policy outcomes than the type of delivery model (Tavistock Institute, 1999). In addition, apart from the experimental Private Sector Lead (PSL) model, which pulled in new private sector agencies, all strategic partnerships tended to involve similar institutions. Typically they included the ES itself, the local authorities, the careers services, further education colleges, the former Training and Enterprises Councils (TECs) in England and Wales (now Learning and Skills Councils) and the Local Enterprise Companies (LECs) in Scotland. In fact, local policy formation was clearly shaped in a top-down fashion, since several local agencies were 'strongly encouraged' to bid to become lead agents:

> There was a lot of political arm-twisting going on in 1997 to make sure that all the public sector agencies got behind New Deal...We had to bid to be the lead agent. It was made very clear that we were to do that. One of two other agencies did try not to have the lead responsibility for certain functions and where there was government funding involved in them, they were very specifically arm-twisted (local authority policy manager).

In many areas, local agencies, which in the past operated independently, have been made to work together on labour market issues. However, the co-ordination and contracting of training provision role represented a new departure for the Employment Service and in general its staff lacked experience in this regard (Education and Employment Committee, 1998; Bennett and Payne, 2000). Moreover, the ES displaced other agencies, especially the TECs, which had previously performed this role. It is perhaps not surprising that its success in creating effective and co-operative partnerships appears to have been highly variable. It has been strongly shaped by the partners' experience of previous networking arrangements, as partnerships appear to have been more effective where the ES managed to co-opt pre-existing networks and relationships between local agencies, and to draw on a prior sense of community involvement:

> There are some partnerships which are more grudging if you like than others. Those are brought together clearly because of a requirement to work together. Successful partnerships depend on how much of a sense of community there is in the locality. If you go to UoD *A*, if you go to one of their partnership meetings, it's like going to see a group of friends. They've known each other for years and they've worked with each other for years...Very very strong, lots of understanding, working very closely together. If you went somewhere like our UoD *B* district, it's a bit more grudging there because as a district it's quite disparate: there's not any big centres of population, the providers are spread out across the district and haven't really needed to work together and may not quite see what there is to gain in having a partnership (ES regional co-ordinator).

Local partnerships tend to be path dependent and are highly reliant on social capital accumulated over time (Considine, 2003). In some cases individual personalities and management styles helped to foster strong working relationships, improve co-ordination, and in some instances, revive interest when signs of 'partnership fatigue' appeared. For example, the manager of the private sector (PS) led partnership in North Tyneside was perceived by his ES partners as helping to achieve a mutual understanding because of his prior knowledge of the ES and its public sector ethos:

> On the cultural side of things, it has gone well because of his understanding which provides the link . . . Certainly that first year before he came, there was a lot of friction; it was, 'Well I mean this and you mean that, and I don't quite understand.' We had lots of meetings about the meaning of things and the interpretation of words. He can translate it into non-ES language for us (manager, Delivery Agent).

More generally, it appears that at the outset of the programme, UoDs were encouraged to include a wide range of agencies within their strategic partnerships. This was meant to allow these organisations to make some input to delivery and it was hoped that this would secure their commitment to the programme. However, as the programme continued the role of such strategic groups often became less clear and the ability to focus the partnership around a shared and commonly understood set of objectives and aims became more difficult. Several responses were common, such as reducing the size of the partnership or appointing a new chairman. More unusually, in the Joint Venture Partnership (JVP) that exists in Birmingham it was decided to establish an operations team with secondees from each of the partner agencies, all based in a single office. Not only did this improve the sense of legitimacy, and built trust among the actors; it also facilitated the routine exchange of information.

Our evidence suggests that the JVP model tended to be the most conducive to joint-working and was certainly seen by many actors as the 'preferred model' (see also Tavistock Institute, 1999). In contrast, other types of partnerships have experienced co-ordination problems. As illustrated above, the PSL model has been characterised by friction between management styles and duplication of efforts due to a sense of rivalry. While the consortium model, if managed and controlled tightly, could increase local commitment, we found that it could equally result in fragmentation as it was not clear whether the Consortium or the Employment Service had the final authority in determining the form and quality of provision. Finally, in the ES-led model, there was evidence of confusion about roles, so that agencies were unclear whether their involvement is purely contractual or also strategic. In this model in particular, agencies express a sense of frustration:

> We're in essence delivering a government programme and we have to make
> sure that we have all the audit documents, and yet we come together in the
> partnership meetings to discuss strategy. Whenever we try to change some-
> thing, we come against this brick wall: 'No this can't be changed.' That has
> been very difficult. New Deal is very much the Employment Service's baby
> and the rest of us have not had as much influence as we would have liked to
> have (local authority policy manager).

For the ES, the advantages of contracting with lead agents for each option
is that the latter will then commit their own resources and expertise, take on
the responsibility of identifying suitable providers from their existing network, and manage all the relevant administration. But this means of sharing
the burden of bureaucracy was understandably resented by the lead agents.
Indeed, heavy paperwork and the incompatibility of IT systems were often
perceived by partners, providers, and even employers, as the major hurdles
to the effectiveness of the New Deal.

Such issues have been highly relevant to the involvement of the voluntary
sector. Most of our selected delivery units evidently recognised that the
diversity of clients, including those from ethnic communities, necessitated
the establishment of an appropriate network of Gateway providers. Some
strategic partnerships either made an early decision to include community
organisations, such as those concerned to support homeless or drug dependent young people. Others engaged in 'capacity-building' exercises with the
intention of integrating small organisations and offering them direct contracts to strengthen the provision available to such groups. However, on the
whole, local partnerships have rarely managed to integrate small organisations that often work part-time and do not have the systems in place to
handle the paperwork related to a New Deal contract. Instead of direct
contracts, the preferred solution has been to devolve the responsibility of
subcontracting to a lead agent, a practice that has had mixed results in
terms of involving small specialist organisations. The Edinburgh UoD,
for example, was the most effective in involving voluntary sector and
community-based organisations. However, this appeared to be largely due
to external factors such as the national commitment of the Scottish um-
brella organisation, the Scottish Council for Voluntary Organisations, to
the New Deal, as well as the existence of a well-developed voluntary sector
network of around 800 organisations.

An alternative, common way of 'capacity building' and involving relevant
organisations was the establishment of working groups (which may dissolve
if necessary), often including groups with expertise on ethnic minority
issues. Those interested organisations that applied to join existing partnerships appear to have been welcomed. One example was the establishment
of a 'virtual' electronic ethnic working group in the Camden/North

Islington UoD, based on email, which helped to raise the profile of ethnic minority groups. Paradoxically, the North Tyneside partners who did not feel that such an ethnic working group was relevant to their area, were directed to consider this option:

> It's really the ES saying we have to have one. We don't have any ethnic minority issues. We've only had 20 people access New Deal from ethnic minorities since it's started, so it's a been a paper exercise. We were just instructed (joint meeting with managers of the ES and the MARI Group, North Tyneside).

Finally and importantly, the formation of partnerships also included the identification of local employers and industries willing to engage in the programme. The degree of success may be gauged from two perspectives: employers' strategic participation through representation on local steering groups, and, secondly, the recruitment of New Deal participants via the subsidised employment option. Employer engagement has generally been disappointing. For instance, varying degrees of employer involvement are evident in local steering groups. Of our case studies, only Birmingham managed to develop an employer coalition of over 20 organisations designed to engage local businesses in the development and promotion of the New Deal.[1] It was chaired by the director of a major UK car manufacturing firm and which included trade union representation as well as Asian and Black business groups. The North Tyneside unit drew on a similar employer coalition, but had to share it with three other UoDs (South Tyneside, Newcastle and Gateshead), since local employers were reluctant to duplicate their involvement. In general, employers located in the more buoyant labour market of Cambridge appeared to be less willing to commit their time, which may be linked to the preponderance of service sector SMEs, often with a less well-developed 'social conscience', in such labour markets. A sense of social responsibility tends to increase in line with a long-standing attachment and loyalty to the local community, which large manufacturing or public sector employers were more likely to display (such as Cadbury in Birmingham).

In general, the level of employer participation in the subsidised employment option was lower than planned. But it is clear that the units located in the least prosperous labour markets, especially Birmingham and North Tyneside (and to a lesser extent Camden/Islington) were more effective than Edinburgh and Cambridge in engaging public sector employers, such as local councils, police forces, health and passenger transport authorities. This outcome may well reflect a stronger collective effort in such localities to offset the lack of labour demand from private sector employers. Overall however, the NDYP has not benefited from strong local government involvement, and local government has provided only a small number of opportunities (Nathan et al., 1998), in contrast to welfare-to-work policies in other European states (Etherington and Jones, 2004). In summary, then, our case studies suggest that the process of partnership building and pooling of organisational resources has been highly variable and constrained, and in this context, it is not surprising that there have also been strong limits to the degree of local adaptation of policy.

## Policy Learning and Adaptation

A second major rationale for localising labour market and welfare policy is argued to be that it facilitates *policy learning and adaptation* (OECD, 1999). Local involvement in the design and management of welfare-to-work policies is claimed to offer the possibility of designing services that are tailored and adapted to different local circumstances (Campbell et al., 1998; Campbell, 2000). It is argued that local actors can draw on local information and intelligence, thereby producing a faster and more accurate identification of needs. Proximity to the local labour market may allow a better appreciation of both client needs and employer expectations. The NDYP was certainly set up with the intention that local partnerships would come to understand clients' needs in their localities and learn about the opportunities for provision offered by local partners and employers, and would then be able to match provision to needs. The official rhetoric envisaged that the local monitoring of the provision should lead to a continuous improvement in its quality.

District managers in each of the case-study UoDs were clearly aware of the potential significance of such local flexibility with regard to their task of identifying and responding to needs. Their aspirations were described by a London ES district manager in the following terms:

> You have to brigade your resources and your knowledge because labour markets are very diverse; networks are very different; each London borough behaves differently. There is no central set of rules, it's just a question of taking your spanner out and tinkering with it. The whole key to this is getting the match right between the labour market and the client. There should be a labour market fit; it's just a question of finding it.

However, in practice the local flexibility that can be exercised was significantly circumscribed. In part, the ability of the programme to respond to local needs was constricted by its dependence on a rigorous contracting process. Local contracts had to be approved and monitored by the regional and national offices of the ES so that, rather than having significant operational discretion, local offices had to seek approval from their regional superiors. In response, some UoDs in the Eastern region and East Midlands used subcontracting by 'strand leaders' to overcome some of the problems of direct ES contracts. According to the Regional New Deal coordinator this was employed to gain greater flexibility:

> If you have ES direct contracts, every time you want to add provision, you have to go through the ES and government contracting procedures, which can actually be a 13-week window to get a new contract up and running. You go through the Treasury rules for contracting, you have proper procurement procedures, you have to invite open tenders, you have briefings, etc., and the whole process can be quite lengthy (ES district manager).

In general, much of the ES Regional Office's role was merely implementing and supervising national decisions and acting as a communication channel between the national decision-makers and the local agencies. A recurrent theme in interviews with staff in both units of delivery and Regional Offices is that they were required to seek approval constantly from higher authorities in the ES hierarchy for any new measures or structures. Some local partners consequently expressed frustration with being trapped in a stale bureaucracy:

> The whole mechanism for contract issuing and management wasn't forward-thinking enough to deal with how New Deal should have been contracted. So although we had a lot of excitement, a lot of enthusiasm, a lot of very good ideas, when they fed through the contract process they came out in little boxes because that's how the mechanism works. That's where the New Deal went wrong. We've ended up with a basic ES contract, so it's not innovative (Development Agency executive officer).

During the roll-out of the programme in the New Deal Pathfinder areas, the Gateway emerged as a source of tension between the ES and other partners and local training providers. Smaller providers, in particular, complained that the ES contracting process was too rigid, protracted and bureaucratic and thus disadvantaged and deterred them relative to large organisations (Education and Employment Committee, 1998). They disliked the use of 'call-off' contracts whereby, even after contacts have been set up and investments of staff-time made, they could be cancelled if there was an insufficient number of client referrals. The low number of client referrals to providers, which made some training schemes uneconomic to run, was a persistent complaint. They also argued that district managers had too little autonomy from regional and national offices to make decisions on contract suitability and design, and pointed out that the compulsory competitive tendering enforced by the ES often meant unique and innovative proposals were undermined by cheaper alternatives. Finally, such organisations also complained that the training allowance per client on the Gateway (about £200) was much too small to allow them to increase the employability of the most disadvantaged young people significantly. Some partners also felt that the four-month period was also far too short to make such clients job-ready. Indeed, some ES officials complained about the 'design rigidities' in the programme, such as the fixed length of the Gateway and the fact that units were castigated for having too many 'over-stayers' if clients remained on the Gateway for more than four months. As one regional manager explained:

> If you don't discover until late in the Gateway that someone needs prepara-tory work before they're ready, there's no point in throwing them in that option or job if they're not ready and fail to make a success of it. There's a lot of problems with on the one hand introducing flexibility and pace which is what the individual requires but then setting up a national model that says, 'These things will happen at these specific times.' The two things don't relate.

In response to such issues an intensive two-week job-search course has been added to the Gateway (see Davies and Irving, 2000) and, as we discuss later, UoDs have started to introduce Gateways more adapted to local labour markets. However, the basic sense of strong design and funding constraints on local flexibility remained. A further example is the design restriction on transfers between options. Once on an option, a client could only move into the subsidised job option; in order to gain access to any other of the options, they would have to go back into benefit claiming, and then start the programme again.

In contrast to previous active labour market programmes, New Deal is meant to be 'client-centred', and one of the most highly praised features of the programme is its adoption of individual case management and needs assessment, in which New Deal personal advisers (NDPAs) can exercise discretion in steering their clients through the programme (Lewis and Walker, 2000). But this is a demanding and challenging role: the failure to identify needs correctly during the Gateway and a lack of awareness of the type of provision available locally could result in participants being directed to the wrong option and eventually being recycled in the pro-gramme. While none of the New Dealers we interviewed were overtly negative about their NDPAs, there was evidence of a dissatisfaction with the degree of support and understanding displayed by the advisers (see also Bryson et al., 2000; Hoogvelt and France, 2000). Reservations about NDPA–client relationships were expressed in all our case-study areas. It was often the case that those clients with higher educational achievement or 'unusual' creative and artistic aspirations tended to perceive NDPAs as unresponsive, if not irrelevant, to their needs. A young person in North London recalled that he had several advisers:

> Some were really good; some were really bad. I had about four, purely because people were leaving, moving offices. I sympathise with them because it was very hard for them to know what to do. I had to put something down that they thought they could help me with. There was a lot of bullying me into doing things I didn't want to do.

We return to the high turnover among NDPAs below. However, as far as responsiveness to clients' needs is concerned a clear difference appeared in the Cambridge UoD, particularly with regard to the subsidised employment option. This was the only UoD where NDPAs had sufficient time to source jobs directly for their clients, without having to limit themselves to vacancies available on the main ES database (LMS) or relying on information filtered down by local marketing officers and regional intelligence units. As a result of lower caseloads, the advisers here also had more time to carry out client visits subsequent to placements. As we have noted already, several studies have reported that policy performance has been better in smaller and more buoyant UoDs, because advisers in such districts develop a more detailed knowledge both of their clients and of local job opportunities. In most UoDs, however, NDPAs did not have the opportunity to network with employers or providers and improve their local knowledge, which poses a major constraint on the matching of clients to employment and training opportunities.

Employers' and providers' perceptions of NDPAs were just as mixed. Employers in Birmingham and North London frequently saw those young people referred to them as having unacceptably low levels of employability. Dismissals from the subsidised employment option were highest in these areas, and employers attributed this to the failure of NDPAs to understand their job specifications and requirements (see Walsh et al., 1999). In addition, Gateway providers often complained about the lack of referrals. They felt that NDPAs were either unaware of the services they offered or simply reluctant to 'let go' of their clients. This issue appeared to be particularly significant in Cambridge and North Tyneside. In the case of Cambridge, the lack of referrals can be explained by the low numbers enrolled on New Deal, which rendered group provision cost-ineffective. In North Tyneside, the explanation lies in the practice of 'vertical integration', which characterises those UoDs led by a private sector agent (see below), whereby the bulk of training provision is kept in-house (Rodger et al., 2000).

The apparent difficulties for NDPAs in properly directing and supporting clients reflected an institutional failure to support their role. While NDPAs were expected to be 'multi-functional', the financial rewards were not commensurate with their responsibilities (Sargeant and Whiteley, 2000). High caseloads (of up to 80 or 90 clients in the busiest Job Centres and delivery units) added to the pressure of working according to predefined administrative targets (placings to submissions ratios), and the requirement to undertake a National Vocational Qualification Level 4 in Guidance, often contributed to high levels of stress and staff turnover. The ES reported that while NDPA turnover rates in northern areas were as low as 4 per cent, they were as high as 27 per cent in some London UoDs (partly reflecting the greater level of alternative job opportunities there) (Education and Employment Committee, 2000).

So, while NDPAs were not in an ideal position to collect and digest all the relevant information, it is also unclear whether the appropriate institutional arrangements for the collection of labour market data and the monitoring of clients' destinations were in place. The record in this particular field has been very weak, especially in Birmingham where the volume of New Dealers was the highest. The New Deal inspection report by the Training Standards Council (2000, para. 53) on this UoD concludes that 'there is no systematic approach to the collection, analysis and use of feedback from clients or employers'. This is echoed by interviews with ES officials and their local partners who, when asked about the constraints they were facing, often mentioned the requirement for enhanced information systems. The Birmingham district manager, for example, remarked:

> When New Deal came in, the information systems that we had were not particularly designed to support the programme, so enhancements have been made to the system; but there's been a limit to what has been possible to provide for the information needs that people have. That will be addressed as we modernise our systems, but we've had to live with what we've got and make the best in setting up additional databases. This is a general problem, but it's just exacerbated by the size of Birmingham; whereas when you're in a smaller delivery unit where you've perhaps got a few hundred employers and a few thousands of clients over the year, it's a more manageable task to load these things and analyse them.

The underlying problem appears to be that while local actors, including NDPAs, were willing to develop local knowledge and be responsive to needs, they were not sufficiently empowered to so. Because local actors operated within a strong framework of centrally set resources and guidelines, learning was slow and the programme functioned on a reactive mode. Future progress clearly needs to be made to enhance the status of advisers, to offer ongoing individual post-placement support and to produce effective tracking systems, as without these the programme will continue to suffer from low levels of job retention.

## Innovation and Experimentation

The third major rationale for local labour market policy is the argument that it can generate a higher rate of policy *innovation and experimentation*. Local improvements can then be diffused more widely. In theory, policy-makers can take more risks at a local level in the awareness that, where innovation goes wrong and leads to mismanagement, the consequences will be locally confined. The key instrument intended to encourage local experimentation in New Deal was the Innovation Fund, which was introduced in November 1999. The Fund totalled £9.5 million over three years and aimed to support pilot projects that would enhance job placement and retention. The major part of the Fund (£5 million) was ring-fenced for projects from the 11 inner city areas with Employer Coalitions. The presence of active local partnerships was essential, since monies from the Innovation Fund were obtained via a joint bidding process.

The early focus of the Fund was on developing flexibility in various operational aspects of the Gateway. In response to its apparent failures, which were to some extent causing clients to drop out, innovation centred on its customisation. Projects included the use of 'licensed centres' where NDPAs could be 'out-stationed' to encourage fuller client participation. They also included initiatives offering one-to-one support, such as the appointment of outreach workers, individual tutors and mentors. In addition, the Innovation Fund has been used to respond to the problem of low geographical mobility among young people in rural areas. Various UoDs in Scotland and Wales and rural parts of England such as Cornwall and Northumberland used geographical mobility funds to support travel-to-work through the payment of grants to individuals. They also implemented car share schemes and purchased scooters, mopeds and bicycles that were loaned to individuals who had secured employment.

In later years the Fund's attention shifted to the development of labour market intermediaries and employer-focused or demand-led strategies (see Fletcher, 2004). Intermediaries are organisations with the capacity to intervene in the labour market to simultaneously improve the labour supply and stimulate the demand for labour (Evans et al., 1999). Intermediary organisations work directly with specific companies or sectors to design customised assessment, education, training and work experience with the aim of ensuring that their clients succeed in the workplace. The aim is that the employers involved get the calibre of recruits that they require. The success of intermediaries hinges on employer involvement, since the Fund encourages them to become partners in projects designed to meet their own recruitment needs, to play a role in the preparation of bids, and match resources in a ratio of 3:1 (they must allocate 25 per cent of the costs).

The overall trend suggests that many local partnerships have failed to secure funds, often as a result of misinformation about the type of ideas and activities eligible for funding. An unpublished ES internal document reveals that out of a total of 600 bids received during the two first rounds in the course of 2000, only 58 were successful. While individual proposals may have involved genuine innovation for individual UoDs, various actors interviewed explained their lack of success by the fact that their project may have been perceived as a replication of a pilot already conducted in another area. The subsequent discouragement and fatigue is not surprising. A regional Employment Service official remarked that the Innovation Fund had sapped the enthusiasm of some individual partnerships.

> The partnership has put in something like 22 bids and had one agreed and they feel that they've gone through an enormous amount of effort with very little return, and they feel discouraged. We have other partnerships that are now saying we won't put any more bids in. And so I do feel that the whole thing has had some problems. We need to be a little bit more focused about how we use the innovation funds, and less bureaucratic because it's been a very bureaucratic system.

A key set of innovations has focused on the targeting of Gateways. All our case-study UoDs had devised, or were in the process of devising, occupational and sectoral Gateways. These tailored Gateways involve educational activities such as job trials and pre-interview preparation. Occupational Gateways are created in conjunction with specific local employers who agree to either ring-fence a certain number of posts for New Dealers or to use New Deal to create new positions, and the example of Nissan's role in Sunderland has been influential. The Birmingham Unit worked with the National Health Service Trust towards the recruitment of nursing cadets, and with Jaguar who recruited apprentice automobile engineers. The local passenger transport authority in Newcastle worked with the North Tyneside strategic partnership to establish new positions of passenger advisers on the Metro network (which was inspired by a similar scheme in the Netherlands). Transco, the gas company, recruited officers for their call centre through an occupational Gateway in Tyneside. Additionally, the North Tyneside partnership also benefited from the presence of Assa, an intermediary that works with Nissan. British Gas co-operated with the ES and a college in North London to recruit gas fitters, as they were experiencing shortages of candidates for these particular posts.

In tighter labour markets the development and piloting of sectoral Gateways has been motivated by the aim of maximising the supply of potential employees. One example is a pilot initiative within the financial service sector in Edinburgh produced by a partnership between ten local banks and insurance companies, the Chamber of Commerce, the ES and the Local Enterprise Company. The recruitment process consisted in sifting clients at various stages: the open day, the training course, and the final interview. On the basis of those who attended the training course, the success rate in securing employment in the sector reached 41 per cent (see Bank of Scotland and Scottish Enterprise Edinburgh and Lothian, 2000). Projects around hospitality, tourism and catering in North London, Birmingham and Edinburgh were less successful and when interviewed, partnership representatives reported that they were attempting to re-engage employers in these sectors through the intermediaries route.

Another area of innovation has been in the options, particularly in the environmental taskforce and full-time education and training options. Those options managed by local councils have been especially innovative. Birmingham City Council introduced various initiatives such as 'City Centre Reps' (tourist guides) (see Helmes, 2004) and a 'cybercycle' project to recycle old computer hardware. The North Tyneside Council developed and delivered a pre-apprenticeship programme with the aim of securing employment for New Dealers with a local construction company. Similarly, the Edinburgh partnership introduced a training course in the construction industry with guaranteed jobs at the end. But in Edinburgh, the training course was delivered by an independent private sector company, which proposed to replicate an intermediary project that it had successfully delivered in Glasgow. In addition, the Edinburgh City Council introduced a project known as 'Deal Me In' in late 2000 (see Lindsay and Sturgeon, 2003). This project's objective was to prepare participants for jobs with the Council and consisted in blending the time on Gateway and the option, thus allowing for the greater personal development of candidates.

There has also been some local experimentation with the use of Intermediate Labour Markets (ILMs), although these have not been funded by the Innovation Fund. ILMs offer sheltered employment and training opportunities similar to those available in the Environmental Task Force and Voluntary Service options, but the major difference is that they pay a wage. ILM projects are either run by the voluntary sector or by local authorities and they usually reflect a genuine and long-standing local concern for social inclusion. Our research suggests that their existence tended to be correlated with local government political affiliation. Out of the five case study areas, only Birmingham and Edinburgh (where Labour had firm control) had ILMs run by the City Councils in which New Deal placements could be made, although the numbers involved have tended to be relatively small. Both had projects around construction and horticulture, 'Birmingham Settlement' and 'Capital Skills'. These were complemented by further

ILMs in the voluntary sector (a café in Birmingham and an energy-saving scheme in Edinburgh).

In comparison to mainstream employment, ILMs appear to be more effective at helping disaffected individuals. They offer a supportive working environment and thus give participants more time to negotiate their transition into work. However, places available in ILMs are strongly related to overall demand in the local labour market. In the context of a buoyant labour market, selection criteria are lowered and the impact is greater since individuals who would not be considered by private sector employers are offered employment opportunities in the ILM. However, given the limited number of New Dealers placed in ILMs, their impact is still relatively small, particularly in the most depressed labour markets, where they are most needed. Cambridge, North Tyneside and Camden/Islington had no ILMs in place, yet the local agents we interviewed recognised their potential merits and often expressed an interest in developing them.

A close look at these examples suggests that much of this innovation, particularly when set up through the Innovation Fund route, followed a centrally suggested course of action. Regular meetings between Regional Office Managers has been a key channel for the diffusion of fashionable ideas, and UoDs appear to have been put under pressure to try for funds. This is especially evident in initiatives that involve mentoring and labour market intermediaries, which are important components of the US welfare-to-work approach and illustrated the importance of transatlantic policy transfer (Dolowitz, 1998). The ES worked closely with the Wildcat Corporation, a New York based intermediary, and presented to New Deal partnerships as a model to emulate. Once again, a top-down formula for policy development prevailed. ES officials consistently stated that innovations were possible in the programme, providing they gained national approval, and in some cases we were told that innovations were discouraged because head office actually preferred a uniform approach. A number of people expressed frustration at constantly having to defend proposed changes to head office. At the same time, this central involvement has meant that the diffusion of best practice appears to have been effective and a number of pilots have been reproduced from one UoD to another. However, this complex mix of national prescription and evaluation with directed replication and local imitation makes it difficult to identify any outstanding and radical innovations facilitated by local decentralisation. One of the reasons, of course, is that innovation does not solely depend on intangible resources such as local networks, partnership working and fundraising expertise. The ability to conceive and implement creative projects such as intermediaries, supportive individual measures and ILMs depends upon the provision and targeting of adequate financial resources.

## Resource Targeting

The final argument often cited in support of local labour market flexibility highlights the possibility of targeting resources more carefully on areas with the highest needs. However, in the case of the NDYP, this has only been possible via the route of match-funding, which involves additional resources being added to New Deal core funding. Indeed, the types of innovative projects discussed were are not funded by the New Deal programme but rather through the European Social Fund (ESF) or the Single Regeneration Budget (SRB). In order to enhance the New Deal through match-funding, partnerships must undergo various bidding procedures. There has been some evidence that the NDYP has linked to pre-existing ESF projects and most partnerships reported that they were preparing new bids. In this respect, the presence of knowledgeable staff in local Government organisations proved a major asset and compensated for the ES lack of experience. Areas traditionally eligible for regeneration and European funding streams tend to be more familiar with the necessary procedures and therefore had an advantage.

However, and more importantly, the possibility of using additional funding such as ESF and SRB is determined less by the expertise of local actors than by the nature of the local labour market. Additional resource attraction is only really feasible in those areas with an established reputation for relative economic disadvantage. A manager of the Cambridge Training and Enterprise Council (TEC) explained:

> This area is classified as not having a need; therefore we do not hit the necessary criteria compared to the Birminghams, the Glasgows, the Londons, the Liverpools; we are disadvantaged by being successful. As a result of that, the people who are at the bottom of the ladder, or not even on the first rung of the ladder, don't stand much of a chance of getting up. And because of the nature of the economy we have, the expectations of employers are far higher than in other parts of the country; so it's a double whammy, in as much as we don't have the funding from central resources and the companies' expectations are higher by and large as well. We are not able because of that to subsidise from other sources of funding when we know that in other parts of the country, ESF, SRB and other things are being used to supplement the New Deal core funding.

Clients living in prosperous labour markets are often assumed to have few problems in securing employment, and pockets of deprivation in more prosperous UoDs may thus be overlooked. Those young people living in such deprived neighbourhoods may well be doubly disadvantaged because their area of residence is ineligible for the type of funding required to develop specialist provision.

A further means of targeting resources was possible in those UoDs that had greater control of their budget and could use unit costs to apply more flexible funding arrangements. In contrast to the majority of ES-led partnerships, the North Tyneside PSL model enjoyed greater levels of autonomy over its budget. In this UoD, for example, it was possible to redirect young people from one specific option to another option without a loss of funding, to offer them an allowance towards clothes, or to 'bend the rules' and use the £750 training allowance, which is supposed to be put towards the acquisition of accredited qualifications, towards driving lessons. UoDs in Wales and Scotland also appeared to have slightly more financial discretion and procedural latitude. In these cases this appeared to reflect the fact that the Regional Offices were not only supervised by the ES in Sheffield but were also reporting to their National Assembly and Parliament respectively. The latter bodies seemed to exert pressure on the New Deal teams to respond to their social inclusion priorities and so created a space in which departures from standard procedures could be considered.

However, central control and the lack of financial discretion in most UoD partnerships was typically justified in terms of accountability in public expenditure (Mulgan, 2000). Further decentralisation and local discretion were rejected on the grounds that they greatly increase the risk of corruption or mismanagement. Privately led partnerships were less subject to public accountability as they operated according to the profit motive. In contrast, Employment Service officials were essentially accountable to national Ministers. For example, a Local Authority manager argued that:

> they will tell you that this all runs back to the Treasury; they are required to be accountable for the public pound, therefore every single piece of money they spend must be followed up by 15 bits of paper to justify the expenditure and that they were stung badly through the previous programmes that they ran for training because there were so many cases of fraud, embezzlement and fiddling, and this time round the Audit Office in England and Wales and the Accounts Commission in Scotland were determined that this would not happen again.

ES district managers enjoyed only a modest degree of discretionary power. Despite the perceived benefits of decentralising welfare-to-work programmes, NDYP has been characterised by an unwillingness on the part of central government to move beyond a certain level of local financial flexibility. However, more recent documents suggest that now this may be changing.

## Work-First Flexibility?

We have argued, then, that, despite the decentralist rhetoric, the NDYP over its first five years actually suffered from a lack of local flexibility. This now appears to have been conceded by the Department for Work and Pensions itself. According to the acting Director, the case for more local flexibility has been won and a more flexible approach giving more power to advisers to decide what is best for clients is preferable to 'the straight-jacketed central planning model with which New Deal was introduced' (cited in Branosky, 2004, p. 18). Indeed the government promises that the reformed New Deal programme to be introduced from 2006 will deliver local solutions to meet individual needs. 'The Government's proposals for the evolution of New Deal are founded on movement towards greater local flexibility and less central prescription, while recognising the need to retain central control over some core elements' (DWP, 2004a, p. 9). The reform plans again stress that the challenge is to combine more flexibility with high national standards in order to tailor programmes but avoid postcode lotteries. The DWP now argues:

> When the New Deal was introduced, there was a need for a high degree of central control to shift the system into a new mode – one that provided more active help and support for JSA claimants to one that offered active help for those who could return to work. As these initial challenges have been overcome, the New Deal has become progressively more flexible, to tackle the challenges of those with multiple barriers, the economically inactive and those in deprived areas (2004a, p. 10).

The trend to greater flexibility has been demonstrated by a number of initiatives, including the introduction of an Adviser Discretion Fund,[2] the introduction of Employment Zones and Action Teams for Jobs in some of the most deprived labour markets and the apparent success of Tailored Pathways, which allow individual advisers to design modular support for individuals (Griffiths et al., 2003a). In particular it is claimed that the 15 Employment Zones set up in April 2000 have demonstrated the value of local flexibility. The Zones targeted long-term unemployed individuals aged over 25, generally those who have been unemployed for over 13 or 18 months depending on the area. Each is allocated a Personal Job Account, as a replacement for the traditional method for claiming and receiving benefits. While funds allocated to this account cannot fall below the amount received from the Job Seekers Allowance, these may be supplemented to respond to individual needs:

There is no fixed amount. The only thing I have to guarantee is that they get the equivalent of their JSA payment. Everything else is at my staff's discretion. It can be as much or as little as required, so with somebody I might invest £5,000 to £10,000; I doubt it, but I could do. Some people cost you nothing because they've been diligent (manager of the Birmingham Employment Zone).

Advisers in the Employment Zones 'had discretion to decide how to help each client and how much money to spend on each client, provided they were able to demonstrate to their local manager a sufficient level of job outcomes' (Hales et al., 2003, p. 9). A larger amount of money can be spent on each client in EZs compared to clients in the ND 25 Plus and so it is perhaps not surprising that the EZs have had somewhat higher job entry rates. The Government has promised to incorporate such flexibility in future reforms. District managers will be have more financial discretion to use a proportion of their employment budgets to meet the needs of the local labour market and to decide what employment programme provision they need. Personal advisers will be given discretion to provide modular provision to individuals irrespective of what benefits they receive, thereby ensuring a flexible menu approach less structured by rules and prescribed routeways (DWP, 2004a).

If such reforms are implemented, then it is probable that the New Deal in future will be better able to respond to individual needs through menu provision. Some of the existing barriers on provision and rigid categorisation of individuals may well be broken down. Furthermore, New Deal partnerships are currently being subsumed into local strategic partnerships and this may produce better co-ordination between local economic needs, development priorities and skills training agencies. But the proposals will also apparently introduce more output focus to expenditure decisions (DWP, 2004a).

The meaning and implications can be gauged from the Employment Zones. The five prototype Employment Zones introduced in 1998 in areas of high unemployment were widely seen as having a progressive social welfarist rationale. Clients participated on a voluntary basis, funding per client was high and Intermediate Labour Markets received strong support. However, the successor, 15 fully fledged EZs, have had a much stronger focus on getting people into jobs and have been much more work-first in their orientation. In the full Employment Zones funding per client was reduced and participation made mandatory, and half of them have been led by profit-making private sector agencies such as Reed and Pertemps. The maximum period of provision was also halved to six months and the requirement to provide training and operate ILMs was removed (Jones and Gray, 2001). Some conclude that the five Prototype Employment Zones were much better able to meet individual needs, especially those of the least employable (Haughton et al., 2000; Jones and Gray, 2001).

Research commissioned by the DWP has shown that the flexibility and work-first focus of the EZs and Tailored Pathways has meant that many advisers have preferred to refer individuals to shorter in-house training rather than to longer external courses that do not lead to immediate job outcomes. The more flexibility that personal advisers have had, the less they have used external provision. Instead, they have tended to opt for in-house provision in order to reduce delivery costs and keep greater control. Many providers of options and other partners have struggled to meet the demands of the work-first flexibility approach:

> Many courses are simply seen as too long, too basic, overly theoretical and insufficiently job focused for an approach whose key objective is to get customers into work and keep them there. Introducing greater flexibility has served to reduce the amount of externally contracted training provision in some areas, diminishing the role of external providers and reducing the capacity of the provider infrastructure in some localities (Griffiths et al., 2003b, p. 21).

Declining referrals to external provision in 'flexible areas' have removed some external suppliers from the provider marketplace, as some have not recontracted and others have gone out of business. 'While the tendency to reduce external contracting may be a legitimate response to deficiencies in the support structure, there is a risk that the capacity of the local provider infrastructure could become compromised' (Griffiths et al., 2003b, p. 23). This would be unfortunate, as clients in some areas already find the range of options and courses open to them to be too restricted. Weakening the training provider infrastructure in disadvantaged areas seems especially perverse when we know that most of these areas suffer from chronic skills and human capital shortages. One response would be to invest and support supplier development in order to improve the quality of their provision, but this has again been ruled out by financial constraints. The reduction of external provision is worrying given that there is evidence that job-entry flexibility is of most benefit to those who are closest to the labour market. It may be that advisers decide not to invest funds in those with little chance of finding work, for instance. The hardest to help typically require longer periods of support and more intensive help than are provided under work-first flexibility approaches (Griffiths et al., 2003b). Thus the type of local flexibility now being introduced will, clearly allow advisers and local managers to have some discretion in meeting goals, but these goals are set at the centre and are rigorously applied through performance indicator measurement and outcome targets. Thus the move to more flexibility will primarily reinforce and extend a frontline flexibility in how to achieve targets and goals that are centrally set. The reforms to policies have clearly been introduced in the context of a strengthened work-first agenda and greater discretion in spending is likely to be conditional on the establishment of rigorous job-entry targets and local performance monitoring. There is some suggestion that the reformed programme will be more demand-led in the sense that it will aim to link to local employers' labour requirements and involve a bigger role for employer coalitions (Browne, 2004b). The New Deal may be moving to more local flexibility but this is a flexibility designed to achieve high job-entry rates; there is no sign that it will extend to allowing local partnerships to decide what their key goals should be and what type of active labour market policy is most relevant to the distinctive problems of their local labour market.

## Conclusions

> Anytime you've got a centralised policy or a centralised programme, the people who design that are reluctant to let it disappear too much into flexibilities in case you corrupt the integral vision of the policy. I understand that but in keeping it so tight sometimes it fails to fit the needs because you can't design at the centre something that is good for every occasion everywhere.
> (ES Assistant Regional Manager)

The spatially uneven development of the NDYP is being driven by two processes. First there is what Jones (2000) has referred to as a 'workfare dialectic', in which the implementation of a national welfare-to-work policy model from above intersects, and is in tension with, locally varying conditions and the search for local flexibility from below. Second there is a 'governance dialectic', in which a specific institutional design – public sector partnerships and contracts with private sector organisations – also incorporates local variation in form and operation. Thus far, the central top-down elements of these twin dialectics have proven dominant. The centralised nature of labour market policy in Britain is clearly proving resistant to change, and the NDYP continues to be delivered by a monolithic centralised institution (Nathan, 2000). Nevertheless, its design has provided some scope for local experimentation and local co-operation, and it is misleading to argue that the entire programme simply represents the imposition of a top-down centralised template. The scale of the problems faced in the most depressed labour markets has in some ways galvanised the New Deal partners to develop some commendable local initiatives. While the limited amount of decentralisation has yielded some of the benefits predicted, there have been major limitations on the extent of local flexibility. This is principally due to the rigid structure of the New Deal programme, particularly the duration and funding of the Gateway and the options. Despite the best efforts of some of the actors involved, learning and adaptation have been restricted because of the way in which the key role of personal advisers has been undersupported and underresourced. The institutional resources for an adequate monitoring of local labour market intelligence and the effectiveness of training and other options appear to be patchy and frequently absent. Partnership-building has undoubtedly occurred in some UoDs but in general it has been greatly weakened by the difficulty of mobilising employers and retaining their commitment. The programme has struggled to build relationships with training providers, especially in the Gateway. Innovation has also been present, but severely constrained by a lack of financial resources. In many cases, innovation has been superficial and less a response to local conditions than a reflection of the competitive pressure for local Units to adopt the latest 'in vogue' initiative.

In summary, however, the paradoxical effect of this restriction of local flexibility has been that the programme has not been able to achieve a uniform level of effectiveness across the country. Indeed, precisely in order to improve the performance of the programme in the more difficult local contexts response, the DWP admits that it requires far greater local discretion and individual case flexibility. However, if the Employment Zones are a guide, the promised flexibility will be based on frontline discretion in the delivery of work-first goals. The precondition for its introduction will be strong performance monitoring and job-entry output targets. In many ways the local flexibility of the New Deal has been contradicted by the other features of the new managerialism that have accompanied its introduction. Thus it illustrates the way in which the rigidities imposed by performance management systems and strict budgetary rules can conflict with the hope that local partnerships will generate new ideas and find new ways of working (see Considine, 2003).

But, local flexibility of itself is not a panacea. While more flexibility at regional and local levels would help the NDYP to better adapt and respond to different specific local circumstances and opportunities, particularly in relation to employers and the voluntary sector, there clearly has to be some overarching common framework of objectives and targets within which implementation takes place. Complete local policy flexibility and autonomy is obviously neither desirable nor permissible. While it can be argued that local discretion and autonomy can make welfare-to-work policies more responsive to local needs and opportunities, there can also be less desirable trade-offs, including an unevenness in programme provision and effectiveness across the country, or a lack of local accountability for the expenditure of public money. A balance therefore has to be struck between local flexibility and a national framework. However, our evidence and the results of other evaluations suggest that, during its first six years or so, the NDYP was unbalanced in the sense that localism was restricted to the local delivery of centrally determined goals and methods. Such an enfeebled localism seemed to act as a substitute for central recognition and response to the severity of the needs encountered by the programme in the most difficult labour markets. In conclusion, the implementation of the New Deal may have marked the introduction of a new type of governance of welfare in the UK, but it has done so, less in the sense of a shift of authority away from the central state to other territorial levels and local agencies and more in the way that the state has restricted and focused its functions and spending by directing, controlling and monitoring the activities and discretion of local administrators (see Cerny and Evans, 2004).

# Chapter Seven

# Conclusions

## The New Deal Policy Paradigm

Hall (1993) has explained the development of post-war British macro-economic policy in terms of changes in 'policy paradigms'. He argues that a policy paradigm is an interpretive framework embedded in the very terminology that policy-makers use to prosecute their programmes and interventions. The influence of a policy paradigm derives not only from the policy measures being implemented but also from the fact that much of it is taken for granted and is unamenable to scrutiny. In Hall's words, 'Policymakers customarily work within a framework of ideas and standards that specifies not only the goals of policy and the kinds of instruments that can be used to attain them, but also the very nature of the problems they are meant to be addressing' (p. 279). Hall draws on Kuhn's (1962) image of scientific paradigms by arguing that only certain types of sociological processes, involving a shift in the locus of expert authority, policy experimentation and failures, result in a contest between competing paradigms and disjuncture in policy:

> Like scientific paradigms, a policy paradigm can be threatened by the appearance of anomalies, namely by developments that are not fully comprehensible, even as puzzles, within the terms of the paradigm. As these accumulate, ad hoc attempts are generally made to stretch the terms of the paradigm to cover them, but this gradually undermines the intellectual coherence and precision of the original paradigm (p. 280).

In this sense, Halls' policy paradigm notion does usefully highlight the significance of the cognitive and ideational frameworks underlying the formulation of policy and the process of policy change (see Capano,

2003; Larsen, 2002; Surel, 2000). It is certainly the case that any given policy paradigm is based – to a greater or lesser extent – on an underlying ideological model of the economy by which it should best be managed (what Jessop [1994] has called an 'accumulation strategy'). In addition, a given policy paradigm requires a corresponding strategy to secure popular support for that model ('hegemonic project'), as well as an appropriate form of 'statecraft' (policy construction and implementation). In these terms, it is possible to highlight and explain the sharp contrasts between Thatcherism and the Keynesian-welfare model it replaced, and between Thatcherism and its successor, New Labour's Third Way. But this analogy between policy paradigms and scientific paradigms tends to suggest that policy change is always an abrupt, revolutionary process, whereas in reality there is often a surprising degree of evolutionary adaptation and incremental adjustment, not only within a particular paradigm itself, but also even between successive and apparently contrasting paradigms: for example, there is much in New Labour's approach to managing the economy and reforming welfare that builds upon elements of Thatcherism.

The initiation and implementation of New Deal policies in Britain since 1997 has been part of an evolving welfare-to-work paradigm that includes making work pay through tax credits and the minimum wage, and an attempt to activate the unemployed and reintegrate them into the labour market. In this policy case, New Labour's infamous pragmatism has been less important than the development of a particular interpretive frame and reasoned strategy, which many would argue is indicative of the wave of neoliberalisation underway in policy regimes in many economically advanced nations.

For example, Jessop (1997, 2002) identifies three main trends associated with this neoliberalisation of policy regimes. The first is the *denationalisation of the state*, with regional and local state forms gaining an enhanced role in governance, regulation and policy responsibility and delivery. The second aspect of neoliberalisation identified by Jessop refers to the *destatisation of the political system*, involving a shift from government to governance, with a decline in the state's direct management and sponsorship of social and economic projects, and an analogous engagement of quasi-state and non-state (private sector) actors. Lastly, according to Jessop, neoliberalisation has involved an *internationalisation of policy regimes*, a process incorporating three main elements: a heightened strategic significance of the international and global contexts within which states now operate, a more significant role for international policy communities and networks, and 'fast' international policy transfer.

The UK's New Deal certainly contains aspects of all three dimensions of Jessop's notion of neoliberalisation. Much of the underlying rationale of the New Deal workfare model was based, at least initially, on the workfare programme developed in the US in the 1990s. Prior to its election to government in 1997, senior figures in the Labour Party had visited the US to see for themselves the operation of local workfare schemes there. Indeed, the New Deal is only one of a number of recent policy initiatives in the UK that have borrowed from ideas and programmes first introduced in the US, another being the Working Families Tax Credits scheme, which has strong similarities to the US. In recent years the ideology of workfare and compulsory activation has diffused across borders and been expressed through variants in several countries. To be sure, national variants, like that in the UK, deviate from the US model, and reflect nationally specific circumstances, institutional set-ups and political orientations. But the key point is that workfare has indeed been a leading example of fast policy transfer (internationalisation), and has become the fashionable mechanism for labour market based welfare reform in several countries around the world (OECD, 1999; Peck and Theodore, 2001). Though they may differ in specific details, and while policies almost everywhere have been evolving, most have been heavily influenced by a common underlying neoliberal focus on supply-side intervention.

Furthermore, the New Deal, like other workfare programmes elsewhere, displays the 'destatisation' referred to by Jessop, in that various quasi- or non-state actors and agencies have been harnessed to carry out policy implementation. The UK's New Deal is built explicitly on a framework of local programme delivery 'partnerships' by which local private sector agencies are contracted to provide services, and in some places to manage, the New Deal programme. This was intended to increase the local flexibility of the programme, enabling it to respond to the particular institutional and private sector labour market agency resources base in each local area. Again, however, the idea of a process of 'destatisation' is vague and can take quite different forms – ranging between corporatist co-ordination and private sector led marketisation – within different national institutional contexts. In the case of the New Deal, while there has been a shift of responsibility to private agencies in some areas, especially through the Employment Zones initiative, and a general use of voluntary and private firms as contractors, in most local labour areas the Employment Service and Job Centre Plus have remained the lead agencies and, as we saw in Chapter 6, have often struggled to engage private firms within local management and planning partnerships. In addition, voluntary agencies have in many places found it difficult to cope with bureaucratic requirements of the programme. 'Destatisation' has often been more of an aspiration rather an accurate description of the NDYP's local mechanics.

As the OECD (1999) emphasises, workfare demonstrates – at least outwardly – the neoliberal trend for state policies to be 'localised', in terms of both design and delivery. Workfare programmes exemplify how economic governance and regulation are being decentralised, if not devolved, from the political centre to the local level. In the UK, from its inception the New Deal was championed by government as representing a new locally based, locally flexible policy initiative, intended to respond more closely to local circumstances and to be more effective in its impact as a result. However, our findings suggest that the degree of 'denationalisation' should not be exaggerated. The degree of local policy devolution and local flexibility of the UK's New Deal have in fact been firmly circumscribed by the political centre, involving the imposition of universal nationwide guidelines, performance targets, close monitoring and financial control. As we argued in Chapter 6, by confining the type of partnership model to a limited number of local-delivery agency models, which are expected to work with centrally set parameters and guidelines, and by imposing strict central targets and heavy output monitoring, the degree of local flexibility allowed has been limited (see also Sunley, Martin and Nativel, 2001). There is clearly a reluctance by the state to devolve too much policy autonomy down to the local level, especially when the policy in question – like the New Deal – is centrally funded and intended to fulfil central policy and ideological goals. Indeed, the same imperative of retaining central political control over decentralised and devolved programmes can be observed with other UK policy initiatives (e.g. in the realms of education, health and the regional development agencies). In the UK, the denationalisation aspect of neoliberalisation has not gone as far as some accounts might suggest.

Our study confirms that the forms of institutional decentralisation characterising nationally specific workfare projects are quite different. Thus while decentralisation in the USA may well involve a competitive struggle between state and local governments to experiment in the degree to which they cut back welfare spending and degrade entitlements (Peck, 2001), decentralisation within Denmark's welfare state has, in some contrast, involved the construction of partnership and consensus between the main social partners at regional and local scales (Etherington and Jones, 2004). In the case of the New Deal paradigm, decentralisation has been centrally directed and carefully managed so that there have been strict limits on both local experimentation and partnership.

Thus, while there have been a series of reforms introduced to the New Deal during the last six years (see selected initiatives, Table 7.1), most of these reforms have been marginal and the policy's key characteristics have been marked by continuity. Many of these reforms have been designed to provide more help to more disadvantaged clients and also to increase the degree of discretion of local over fairly contained and specific measures. Indeed, most of these reforms (with the exception of StepUP) have been about intensifying the job search and supply-side help provided to targeted

**Table 7.1**  New Deal reforms and related initiatives

| Schemes and Initiatives | Content |
| --- | --- |
| Employment Zones, 2000 | Private and voluntary contractors provide action plan, job search and personalised job account to assist long-term unemployed, lone parents and NDYP returners find work, 13 areas |
| Action Teams for Jobs, 2000 | Outreach teams provide funds or support to remove barriers to work and access jobs within travelling distance, in 65 deprived wards |
| Tailored Pathways, 2002 | Personal advisers offer more flexible and modular support, combining different New Deal Options |
| Employment Retention and Advancement, 2003 | New Deal 25 Plus and NDLP customers given dedicated advisor and retention and training bonuses, 6 JCP Districts. |
| StepUP, 2003 | Guaranteed full-time job and support for up to 50 weeks for those who remain unemployed six months after completing a New Deal option or the IAP in New Deal 25 Plus, 20 areas |
| Adviser Discretion Fund, 2003 | Discretionary one-off payments by Advisers (up to £300) to overcome barriers to work |
| Progress2work and progress2work linkup, 2003 | Training, education and specialist work advice for drug and alcohol misusers, ex-offenders, and homeless |
| Pathways to Work, 2003 | Adviser support, access to employment programmes, plus in-work credits for those on incapacity benefits in 7 areas |
| Working Neighbourhood Pilots, 2004 | 12 pilots in high worklessness areas, accelerated access, interviews, local funds, retention payments |

groups, usually within defined areas. While there has been a recurrent use of local pilots as a means of policy experimentation, thereby producing a mushrooming of place-based, supply-side schemes, our impression is that local experimentation has been centrally steered and fairly marginal to the development of policy. It has, of course, conveniently reinforced the impression that there are peculiar and exceptional problem areas where special measures are needed and are being supplied. Moreover, the large number of small and targeted initiatives have only heightened the confusing fragmentation of policy initiatives on local economic employment and development. But, in the UK, it is not convincing to argue that the introduction of workfare has produced a geographical process of competitive welfare degradation and local unravelling of uniform welfare rights. Conversely, neither has local experimentation unleashed a strong progressive logic of local partnership building and co-operation in labour market policy delivery. Instead, it is clear that to date there has been no radical break with the supply-side paradigm that laid the foundations for the structural features of the New Deal. The Government's recent promise to build on New Deal by introducing a menu-based approach for client provision and more local flexibility from 2006 is likely to increase local manager discretion significantly (DWP, 2004a). It may prove to be a more significant policy change, introducing more radical local decentralisation in welfare practices and rights, although it remains to be seen how and how much of this agenda will actually be implemented. However, the welfare-to-work paradigm will clearly continue to frame any decentralisation, so that the result is likely to be a contradiction or tension between aiming to allow local flexibility, so that local managers and agencies can respond to their local problems and, at the same time, predetermining the understanding of these local problems so that responses are channelled into intensive supply-side employability-focused activities.

We have argued throughout that New Labour's workfare project is ambiguous. Is it neoliberalism by another name? Is it a Third Way social-democratic modernisation of welfare, addressing jointly problems of social exclusion and social responsibility? Is it dominated by communitarian impulses that focus attention on social integration rather than social inequality? Is it American, or European? A case can be, and has been, made for all of these (and other) views (e.g. Burden, Cooper and Petrie, 2000; Clarke, Gewirtz and Mclaughlin, 2000; Glennester, 1999; Hay, 1999, 2002; Powell, 2002). There is no doubt that New Labour's 'modernisation' approach to public policy in general, and to welfare more specifically, is a neoliberal orientated project. But the elements and emphases are different from its US counterpart, even the Clinton New Democrat form; they are also different from the European social model (Clarke, 2004; Liebfried, 2000; Taylor Gooby, 2001).

Applied to the issues of unemployment and welfare, New Labour's Third Way credo stresses the importance of restoring individual responsibility and the work ethic, and the importance of mutual rights and responsibilities. It is not simply a socially authoritarian and paternalist form of 'roll-out neoliberalism' (see Peck and Tickell, 2002), as it is also combined with a concern about market failure and a genuine commitment to reducing social exclusion and poverty. The guiding belief is that in the past the state has subsidised inactivity and unemployment without requiring and enabling the unemployed to participate in schemes to improve and restore their employability. This has encouraged worklessness rather than work, and has thereby helped to create a 'dependency-culture' problem. The solution is therefore to make benefit receipt conditional on job search and job acceptance and to increase simultaneously the financial incentives of moving into paid employment. The replacement of benefit offices and job centres with the 'one-stop' Job Centre Plus from April 2002 was a symbolic step, as it not only makes a work-focused interview mandatory but also extends the presumption that conditionality should apply to all working-age benefit claimants, excepting only those who are physically unable to work. The core concept underlying activation is that of a 'contract' between the state and each individual client. In this imagined contract, those who can work should accept their responsibility to work and, in return, the state will provide an individualized service to provide help with employability, job search and training. For the majority, the outcome is to shift social citizenship away from being an unconditional status or right towards being a contract that carries mutual benefits and obligations (Plant, 2003; Handler, 2004).

This paradigm is clearly something of a hybrid. It combines a strong neoliberal work-first orientation with a Third Way paternalistic desire to reduce social exclusion and poverty by providing some form of equality of opportunity. Thus it relies on neoliberal workfare but is tempered by a genuine concern to rehabilitate and support the unemployed and inactive. The strategy is also intended to reduce public spending on unemployment benefits, releasing funds for other social priorities, including health, education and alleviating child poverty (Jessop, 2003). As several authors have pointed out, the paradigm is marked by contradictions. There is an apparent conflict between the claim that work-first strategies reduce wage inflation by increasing the labour supply at the lower end of the labour market, and the claim that one of the key aims is to make work pay by improving the financial rewards from entry-level employment. On the basis of this, of course, radical critics tend to dismiss the social inclusion and poverty reduction goals and argue that the real aim of policy is to serve economic purposes by supporting low-wage employment growth in a neoliberal labour market, and cutting the welfare bill (Ferguson, 2002; Peck and Theodore, 2000a, 2000b).

Moreover, as we have seen, there is also a contradiction between the bureaucratic regulation, performance monitoring and managerization of the new policies, and the claims that local agencies have discretion to tailor policies to the local level, just as individual case workers should be able to exercise creative discretion to provide individual clients with tailored support. Furthermore, there is also a tension between the necessary selectivity, breakdown of trust and exclusion implied by conditionality enforced by sanctions, and the aim of increasing social inclusion. The implication of these problems may well be that the core idea of a contract between individual clients and the state can be only imperfectly realised in practice. As Handler (2004) argues, the unemployed are in a weak and dependent position, with many effective limitations on their choices, so that they are not able to enter a contracting relationship as free and equal partners. As we have discussed, there are various bureaucratic and financial constraints on the NDYP that justify a certain scepticism concerning its responsiveness to individual clients' needs. Bureaucratic monitoring and the pressure on agencies to meet work-first targets inevitably creates the tendency to focus on the easiest to help, at the expense of the assessment and resolution of the problems of the harder to help. In addition, as Plant (2003) argues, this shift towards a contribution and obligation view of citizenship clearly 'does require one to be optimistic about the future of jobs and the buoyancy of the labour market, because if the jobs are not there, then the government would be willing the end and not the means' (p. 163). The notion of a contract depends on a balance of responsibilities with opportunities and this is precisely where the issue of the geography of the New Deal programmes becomes crucial.

## Geographical 'Anomalies' in the Performance of the New Deal

There is little doubt that the prevailing welfare-to-work paradigm has received a good deal of reinforcement by the experience of the New Deal, and especially the youth component of the programme. The positive feedback to evaluations of the personal case manager approach, together with the respectable aggregate outcomes and fiscal savings of the NDYP in the context of a buoyant national labour market, have reinforced the commitment to a work-first strategy. As Walker and Wiseman claim, 'To date, New Deal policies have enjoyed (and possibly contributed to) the most favourable of economic conditions' (2003, p. 24). But even within this context our results have demonstrated that, as theories of neoliberal workfare rightly predict, the work-first orientation of the New Deal has meant that its effectiveness has been highly contingent on the status of local labour demand. It has worked best and hit more of its targets in areas of buoyant employment growth. In contrast, the limitations of the work-first strategy laid out in preceding chapters (particularly its tendency to produce recycling as individuals enter and exit insecure and precarious jobs, and its inability to develop human capital) have been most clearly visible in depressed local labour markets. The inferior outcomes in these more difficult situations cannot be explained solely in terms of the quality of local labour or variations in the effectiveness and energy of local agencies, and in this sense they are anomalies that are not fully comprehensible within the terms of the political-ideological paradigm underpinning the New Deal.

There has been little serious examination of these geographical anomalies in official discussions of the New Deal, and it is striking that certain forms of evidence identifying problems more compatible with the prevailing paradigm have received much more attention. As Brodkin and Kaufman (1998) argue, what is crucial in welfare reform is the relationship between evidence and what they term 'ordinary knowledge', or the commonly held beliefs that constitute a general body of assumptions among policy-makers. In their view:

> Findings consistent with prevailing political agendas and ordinary knowledge seem to be heard differently from those that are inconsistent. For politics to hear – or seriously incorporate – contrary evidence requires a higher burden of proof, one sufficiently compelling to overcome ordinary knowledge and to contradict the implicit causal stories embedded in it (p. 17).

The New Deal paradigm has clearly 'heard' the problem of those individuals with multiple barriers to work, but has been much less attentive to local variations, whose complex causation leaves enough room for ambiguity and the survival of conflicting causal stories. There are some signs of paradigm stretch however, and it is now admitted that there are residual localised pockets of worklessness in very small defined areas. But the problem of these areas is not seen to be a lack of jobs; rather, it is a problem of cultures of worklessness. In the words of Des Browne (2004, p. 1) (Minister for Work), 'The problem is not a lack of jobs, it is about making people believe they really can work again.' A recent joint publication by the Treasury, Office of the Deputy Prime Minister and Department of Trade and Industry rehearses the less than fully convincing argument:

> In theory, relatively low employment rates in certain areas could reflect low demand for labour by employers. In reality, however, differences in employment rates do not appear to be due to a lack of jobs. Most non-working adults live in cities but every city in the UK has more jobs than it has residents in work. There are also vacancies in every region of the country and the number of people chasing each vacancy declined sharply over the past decade. Far from there being no jobs available, there are in fact jobs available in all regions (2004, para. 2.81).

In general, the official view on geography and the NDYP has been that labour mobility can cope with any localised pockets of surplus labour. For instance, as Table 7.1 indicates, in 2000 the Government set up a total of 65 local Action Teams for Jobs in urban depressed wards in order to tackle local barriers to work. However, this scheme seems to be predicated on the claim that there are more than adequate numbers of vacancies available and is therefore aimed at reducing the mismatches and improving the bridges between unemployed people and vacant jobs. For example, the Action Team website states that certain skills may no longer be in demand, but repeats the mantra that 'There is rarely a lack of jobs within travelling distance of disadvantaged areas and there are plenty of unfilled vacancies in the places where jobless people live' (DfEE, 2000, p. 1). In fact, the evidence for this claim is dubious and neglects some broad patterns in the geography of worklessness. As we noted in Chapter 2, there is a regional pattern with high levels of inactivity in the north-west, north-east, South Yorkshire and inner London (Burkitt and Hughes, 2004). Indeed, the Social Exclusion Unit's (2004) research report on deprived areas appears to offer a slightly revised take on understanding local concentrations of worklessness, as it admits that changes in the nature of location of employment do play a contributory role, together with areas effects and residential sorting by the housing system. The report pays homage to earlier Government assertions that there are no local jobs gaps, by claiming that 'The Government has consistently concluded that, in the majority of places, there is no shortage of job opportunities' (para 4.5). But it then points out that while concentrations of worklessness exist in local authorities with high and low levels of vacancies relative to the numbers of unemployed resident in the area, two-thirds of people in such concentrations live in local authority districts with above the national average ratio of unemployed people to Job Centre vacancies. Moreover, concentrations of worklessness are twice as likely to be found in local authorities experiencing a fall in workplaces between 1998 and 2001. The report concludes that 'Taken together, these figures suggest that a lack of accessible jobs does contribute to concentrations of workplaces in some places' (para 4.13).

So why is it that there has been so much denial of the importance of geographies of job opportunity and employment demand? The first reason is that this conflicts with the supply-side diagnosis of the unemployment problem. The Treasury and DWP have repeatedly argued that all areas have available vacancies within feasible travelling distance, that there are more jobs than residents in all cities and that the availability of vacancies and jobs explains only a small amount of variation in employment and unemployment rates (HM Treasury and DWP, 2003). But there are real doubts about the accuracy of local vacancy statistics. The level of vacancies used in the UK is derived in a blunt and imprecise manner by multiplying recorded vacancies by three, when this ratio itself is likely to vary across different local labour markets. It remains the case that even if the workforce vacancy rates are similar across localities, those with higher unemployment rates continue to have more unemployed people for each vacancy (Webster, 2000). The simple fact that cities have more jobs than workers tells us more about the ·importance of commuting than it does about the real level of demand for different labour force segments within travel-to-work areas. As the Social Exclusion Unit (2004) admits (rather more candidly because of these measurement difficulties, the importance of the quality of jobs and interactions between demand and supply), it is impossible to say categorically how many areas have 'enough' jobs (para 4.6).

In fact, a recent comprehensive econometric study on the relationships between vacancies and unemployment at the local level reports that there exists a statistically significant negatively sloped relationship between total vacancies and local unemployment across local areas, which is consistent with an underlying 'UV' (unemployment and vacancy) or 'Beveridge' curve (Dickerson, 2003). Not surprisingly, areas with high rates of unemployment also tend to have low vacancy rates, and vice versa (Figure 7.1). Local variations in labour demand conditions obviously do exist, and indeed are remarkably persistent. Moreover, if anything, such studies are likely to understate the position of the UV curve seriously across localities, and hence the significance of local deficiencies in labour demand. It is estimated that there are presently some 6 million or more people of working age who

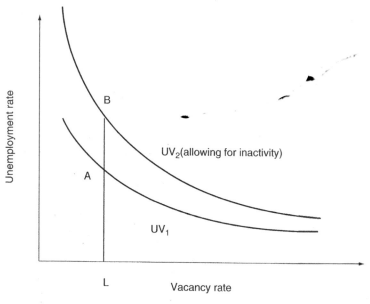

**Figure 7.1**   The relationship between unemployment and vacancies across local labour markets.

are out of the labour market (neither employed nor officially unemployed). Of this figure around 2.75 million are on incapacity benefit. As we saw in Chapter 2, inactivity rates vary dramatically between local areas across Britain (from about 6 per cent to almost 35 per cent). Moreover, local labour markets with high unemployment rates also tend to be those with the highest inactivity rates. Assuming – not unrealistically – that a high proportion of the economically inactive could in fact work, and probably would like to work if suitable jobs were available, this means that local variations in unemployment are much higher than suggested by official rates, so that in many localities (such as L in Figure 7.1) a given vacancy rate is associated with a higher real jobless rate than standard unemployment figures suggest: in other words the 'real' UV curve is more like $UV_2$ than $UV_1$, and high unemployment areas in fact have larger jobs gaps than official data imply.

As we also suggested in Chapter 2, the fact that inactivity rates tend to be higher in localities with high unemployment indicates how local labour demand conditions cannot just determine the level of employment, but also shape the nature of local labour supply, and its employability. The existence of persistent high local unemployment – as in certain inner city areas, or in older run-down industrial zones – can give rise to a whole range of mutually reinforcing social effects that make for a lower-skilled, less motivated, demoralised and socially (and residentially) excluded workforce (many of whom may drop out of the labour force altogether), which finds it additionally difficult to acquire jobs in what is already a generally depressed local labour market. In short, many of the supply-side problems (such as low skill, lack of motivation) stressed by neoliberal interpretations of unemployment, can be the direct and indirect products of a lack of local labour demand. Localities can this become trapped in a sort of low employment to low skill equilibrium.

What also matters, of course, is not simply the nature of level of local employment, but also the nature of the jobs available. This issue also seems to have been given insufficient attention by the Government in discussions of the New Deal. For example, Forth et al. (2002) show that London's deteriorating unemployment performance in the 1990s was due to a slower rate of jobs growth than the rest of the country, a substantial outmigration of jobs, and to a high workplace closure rate. These problems were exacerbated by a relatively poor performance in the growth of skills among Londoners so that the rise in demand for high-skilled workers has increasingly been met by skilled workers commuting into the capital, leaving low- and non-skilled members of London's population stranded without work in rundown, and socially excluded residential neighbourhoods. Another econometric study commissioned by the DWP on the determinants of Job Centre Plus performance also finds that the most significant determinant of JCP district performance, as measured by job entry for JSA claimants, is the level of local demand and vacancies, the second most important factor is the budget per client and the third is the index of multiple deprivation (GHK et al., 2004).

However, the question of local demand has not only been downplayed because of a particular ideological position on the causes of unemployment. Perhaps more importantly, the idea of a local jobs gap appears to dilute some of the core moral and ethical imperatives of the paradigm. The key normative idea here is the insistence that the unemployed for whatever reason have lost sight of their responsibility and duty to return to work. Admitting that there are shortages of vacancies in some areas risks undermining the pressure on individuals to find a job and weakens the work ethic argument by legitimating their unemployment. Fundamentally, the uneven geography of the New Deal for Young People raises a question mark over the core idea of a new contract between each client and the state. For the new contract idea implies that responsibilities should be accepted in exchange for opportunities and a modernised welfare service providing individually sensitive support and help.

The uneven geography of labour market demand seriously limits the delivery of both of these promises. It does so, first, because as we have seen, the concentration of large numbers of clients in depressed local areas places great pressure on the time and abilities of welfare staff in these localities, so that there is a tendency in such places for the relations with clients to become perfunctory and bureaucratic. Second, the relative lack of secure job opportunities and scope for career development threatens the delivery of real opportunities in structurally depressed labour markets. The result is an uneasy imbalance between the belief in uniform and universal responsibilities on the part of working-age individuals on the one hand, and geographically variable job opportunities on the other. As Plant (2003) comments, what this means is that the government is emphasising contribution and reciprocity as a central condition of citizenship without being able to guarantee that society can keep its side of the bargain. 'The logic of the government's position that work is the passport to economic and social and economic citizenship and the way out of social exclusion is for the State to become the employer of last resort, or at least the work funder of last resort, where ordinary labour markets do not produce the jobs' (Plant, 2003, p. 163). Certainly, if this geographical inequity is to be addressed, then there is a need to reform the New Deal programmes to significantly increase the opportunities they deliver in depressed local labour markets and guarantee jobs after a lengthy period of unsuccessful job search. Only by doing this, we would suggest, can the aim of a genuinely contractual and mutually beneficial bargain be delivered in a spatially uniform way.

## Increasing Local Opportunities

The lack of opportunities offered by the New Deal in distressed local labour markets ultimately reflects the UK's relatively low level of spending on activation measures. The NDYP has been marked by a substantial financial underspend, in that as a result of favourable economic conditions, the total expenditure is now estimated to be less than half of the initial budget. Indeed, while spending per client has been higher than in many American workfare schemes, the gap between the UK's spending on active labour market measures and that of other European countries has widened under the New Deal era (Clasen and Clegg, 2003). One of the key difficulties faced by the government here is that the moralising paternalistic tone of the New Deal's emphasis on restoring the work ethic may be popular, but it makes it very difficult then to justify spending large amounts on active labour market policy for the workless (Clasen and Clegg, 2003). If this ideological obstacle were to be lowered and spending on activation were to be given higher priority, then some of the deficiencies in the current policy could be addressed.

In the first place, the human capital side of the New Deal's supply-side focus has been relatively weak. This is despite the fact that many deprived areas are known to have endemic skills shortages and high proportions of the workforce in deprived areas have no qualifications – as a whole nearly 50 per cent of the UK population has a low numeracy level, defined as being equivalent to the numeracy of an 11 year old (Tamkin et al., 2005). At the same time many tight labour markets suffer from skills shortages. The lack of a real contribution to training is clearly linked to the problems of low job retention and workforce recycling under welfare-to-work. For instance, a recent evaluation of the jobs attained in the Employment Zones reported:

> On the evidence of this study, it was relatively unusual for recruits (whether from the main programme or otherwise) to receive training in the job they were doing. The jobs were predominantly low-skilled and low-paid and were inherently insecure even when not known at the outset to be temporary in nature. All-in-all there were probably few opportunities for those in work to progress towards better jobs within the organisations where they obtained their first jobs. If they were to achieve job progression, they would usually have to change to a different organisation, and that begs the question of where they would find help to achieve this progression (Hales et al., 2003, p. 97).

Perhaps even more disconcerting was the finding that when these clients in the EZs lost their jobs they reverted to unemployment rather than demonstrating higher employability (Hales et al., 2003, p. 97). Despite the severity of problems, training and human capital approaches have been heavily criticised for not improving job outcomes and for failing to raise the earnings of clients. Layard (2004) argues that evaluations suggest that training schemes may reduce cumulative earnings by delaying entry to work, 'So the best thing for an unemployed person is to get them into work fast' (p. 6). As we have seen, the education and training dimensions of the NDYP have likewise been criticised in terms of their limited effects in getting people into jobs and for not delivering training relevant to local opportunities. However, the longer-term benefits offered by training pro- grammes are still uncertain and very hard to measure. What is clear, though, is that training programmes are most effective when they are close to the labour market and integrated with actual employment (see Meijers and Riele, 2004). This suggests that supply-side training pro- grammes are unlikely to be sufficient on their own, as in many parts of the UK a low-skills equilibrium exists in which employers specialise in low- skill sectors and strategies that demand only a low-skilled and contingent, disposable labour force. Skills shortages have been found to be higher in more dynamic areas of employment growth (Green and Owen, 2003).

A more effective solution to providing skills upgrading may well be to develop a demand-side approach that encourages employers to develop more high-productivity, high-quality strategies (Payne and Keep, 2003). Such a demand-side skills policy has been discussed in recent years, but does not sit comfortably with the current neoliberal policy emphasis. But some of the deficiencies of current skills training for the young unemployed could be addressed through local programmes that involve employers and that subsidise and encourage firms to provide opportunities for genuine training and progression, as part of a productivity development strategy. In this way skills policy might ultimately become more preventa- tive, rather than simply reintegrative, by improving job progression and reducing labour turnover. The Government has introduced Employer Training Pilots, which are run by Local Skills Councils in 18 areas. These offer subsidies and free training courses to encourage greater training of the low skilled up to NVQ Level 2 (see Hillage and Mitchell, 2003). However, while these Pilots may engage low-skilled employees in qualification-based training and so start to tackle some social inclusion objectives, they do not appear so far to be having a major impact on the nature of skills provision in these areas (Tamkin et al., 2005). Tamkin et al. (2005) conclude that tackling low skills equilibria may require 'wider policy interventions such as greater levels of business support and/or placing greater imperatives on business to raise their levels of productivity and therefore skills' (p. 32). While there has been much discussion of the desirability of a demand-side skills policy of this kind, there are few practical proposals for its implemen- tation (Lloyd and Payne, 2002a). Finegold (1999) suggests that the UK should build on existing high-skill sectors by to emulating the high-skill clusters or ecosystems found in the US. While this proposal rightly highlights the heterogeneity and diversity hidden by national perspectives, as a strategy it is likely to widen territorial disparities in economic well-being and some of the wealth created would have to be transferred to depressed areas where non-employment is most problematic. Keep (2000) proposes that moving towards a higher-skill trajectory demands greater redistribution of income to block off markets for cheap and standardised products and services, and argues that UK companies will need access to more patient, longer-term capital in order to allow them to implement job redesign and high performance practices. According to some, even these institutional and policy measures will not be radical enough to shift the UK economy away from its predominantly low-skill specialisation towards a higher skill path. Lloyd and Payne (2002b, 2003), for example, argue that such a step-change would require a fundamental political realignment and departure from the UK's stockholder model of Anglo-Saxon liberal capitalism. But, as ever, it is important that the case for such a radical transformation does not preclude incremental reform. There is an obvious need for research into the pragmatic steps that can be adopted in low-skill local economies to implement demand-side skills policies by encouraging, and providing incentives to, employers to develop higher-skill strategies.

Even in their absence, it is still important that measures are designed and implemented to improve job retention. For instance a longer-term approach based on a 'local attachment agency' or 'employment and employability' agency focusing not just on job search and job entry, but also on helping young people to build more sustainable career paths through post-placement support and retention, and career advice and skills development, would offer clear benefits (Local Government Association, 2001; Nathan, 2000; see Peck and Theodore, 2000b, for a US example). Such an agency could use an attachment account (similar to the Personal Accounts used in the Employment Zones) to further employability (Evans, Nathan and Simmonds, 1999). But this would imply a radical change in the length of intervention, with a need for long-term relationships to be built up between personal advisers, clients, employers and other local partners. Such an agency could also be a means of ensuring a more effective co-ordination between all the key agencies at local level and of increasing the (highly limited) role of local government in the implementation of local employment policy. A key aim of such an agency would be to ensure that public expenditure is, wherever possible, retained within the local economy and used to foster local employment growth (Local Government Association, 2001). Etherington and Jones (2004) point out that the far greater involvement of local government in Danish local labour market policy has helped to sustain labour demand in areas affected by significant economic restructuring.

One of the means of doing so is through the provision of forms of transitional labour market. Transitional labour market schemes can be based on either a 'jobs pool' approach where local employers reserve a number of temporary jobs for the unemployed, or on the provision of employment in community businesses. Such labour markets typically offer bridging temporary work experience and other support to the long-term unemployed in order to improve their motivation, confidence and employability. They include Intermediate Labour Markets (ILMs), which are local initiatives providing temporary work experience in a genuine work environment with additional support and training to accompany the transition to work (Marshall and MacFarlane, 2000; Finn and Simonds, 2003). ILMs also attempt to contribute to local community regeneration by providing services that improve aspects of physical, housing or social environments. By supporting local regeneration they are intended to have a preventative as well as reintegrative function. They are usually devised and managed by local partnerships that draw and integrate several funding streams. In this way ILMs should be distinguished from the generic concept of transitional employment programmes (TEPS), which refers to temporary wage-based initiatives for the long-term unemployed, where the jobs provided may be in the public sector and where programmes are often national rather than local.

The boundaries between job-creation schemes and ILMs are not clear cut, but the difference is usually one of scale and the types of agencies involved in design and delivery (Meager and Evans, 1998). TEPs also make more extensive use of private sector job placements with the aim of offering temporary relevant and realistic work. Several authors have argued that the fuller provision of such transitional labour markets would help to remedy some of the weaknesses of the New Deal and would represent a proper response to the ethical problem of the geographically uneven opportunity to find work. White (2000), for example, argues that 'an ethical commitment to provide universal access to productive work can only be met by the construction of an institutionalized sector of intermediate employment'(p. 295). Finn (2003) also points out that, in the light of the variable job-entry rates seen under New Deal,

> If the UK is to create something like the ethical employment policy of countries such as Sweden and Denmark, then it will need to create a viable intermediate employment sector for the hardest to help. This does not necessitate a return to the 'make work' schemes associated with old-style job creation programmes but a harnessing of the potential of Intermediate Labour Market projects that have developed in high unemployment areas to deliver transitional waged jobs for the long term jobless (pp. 721–722).

As this suggests, ILM initiatives have grown significantly in recent years and now exist in many of the most depressed local labour markets in the UK.

They have typically been initiated by local groups and have been heavily reliant on funding from the European Structural Funds, as well as the New Deal and Single Regeneration Budget (Marshall and MacFarlane, 2000). Existing evidence suggests, not surprisingly, that ILMs are more expensive per client than more standard labour market provision although there is some debate on their exact costs.[1] There is also uncertainty about the beneficial impacts of ILMs, as many studies have been anecdotal and the initiatives are designed to meet a diverse set of goals. However, recent surveys have highlighted high rates of job entry and beneficial results in terms of improved employability, although in some cases this may be because voluntary participation produces a selection of more motivated clients. Much of course depends on organisation and the precise combination of wages, training and support. The more extensive ILM provision becomes, the more it is likely to run into problems of displacement, and there is uncertainty over the scale at which ILMs can be mobilised. They are, therefore, especially well suited to targeting in deprived geographical areas where the services they deliver can make a genuine contribution to supporting community regeneration. A large number of British ILMs have already been involved in New Deal option provision and Finn and Simmonds (2003) estimate that about 10 per cent of New Deal 18–24 ETF and Voluntary Sector and New Deal 25 Plus work experience options have been delivered by ILMs. However, Finn and Simmonds conclude that the Government's current attitude to ILMs can best be described as 'neutral'. ILM provision has been constrained by the bureaucratic requirements and rigidities of New Deal funding. ILM participation apparently needs to be up to a year to deliver significant gains in employability, which of course is much longer than a New Deal option. According to Finn and Simmonds (2003):

> While the survey has shown that many ILMs have successfully worked with the New Deal, it is our view that there is also a persistent tension between the design and structure and rules of a national programme and locally devised ILMs. The extent of flexibility permitted at the local level to 'bend' national programmes is therefore critical to the ability of local partners to operate ILMs and other local labour market initiatives (p. 94).

Providing better support for ILMs would require greater local flexibility and lighter administrative requirements that encourage local strategic partnerships to devise programmes that respond to distinctive local problems and opportunities. Whether this actually occurs or not will depend on the nature and degree of local discretion permitted in the forthcoming reformed New Deal.

However, it is clear that policy has tried to learn from some of the advantages of ILMs. Indeed, in recent years many governments have implemented temporary employment programmes that provide a combination of wages, training and other support. The current wave of TEPs is somewhat different from earlier countercyclical public employment schemes that were found to be least effective for the hardest to help. Recent TEPs have frequently deliberately targeted the long-term unemployed, they have often involved indirect job creation through subsidies, and operate to supply a job guarantee after a specified period of unemployment. For example, several other European countries have extended the opportunities available to young people and the long-term unemployed (Gardin and Laville, 2000), either via the public sector route such as in Sweden and Denmark, or through intermediary institutions such as social co-operatives in Italy (OECD, 1999), or more recently through employment and training companies in the eastern region of Germany (see Nativel, 2004). Another recent example is provided by the French *Nouveaux Services-Emplois Jeunes* introduced in late 1997 and targeting the same age group as the NDYP. The French programme aims to develop new jobs and careers in line with the potential demand for local 'proximity' services. Contracts are offered for a period of five years with the aim of achieving *pérennisation* (sustainability) (Seuret, 2001). There are no doubt economic dangers involved in such schemes (Calmfors, 1995), so they must be carefully managed to target the least employable and to ensure continued job search and wage flexibility.

The UK government is experimenting with a TEP 'jobs pool' approach through its StepUP pilots. StepUP offers a guaranteed full-time job and support for up to 50 weeks for those in twenty local areas[2] who remain unemployed for six months after completing a New Deal option or the Intensive Activity Period in New Deal 25 Plus. The jobs are sourced by a local managing agent from employers in the private, public and voluntary sectors. A support worker is responsible for providing advice and guidance to each client throughout the contract. Employers are paid a wage subsidy for 50 weeks of at least the minimum wage as well as a fee to cover their additional costs. Participation by the unemployed is mandatory and sanctions may be applied to those who refuse to accept a job. The Pilots are still in an early stage but evaluation to date suggests that they have been positively received by participants, providing labour market discipline and positive motivation (Bivand et al., 2004). Private employer participation has apparently also been high largely because of the generosity of the subsidy offered. It is too early to judge what the job outcomes will be and it is currently unknown how many participants will be offered unsubsidised jobs at the end of the 50-week period. Once again, however, there are indications that the results of the programme will be dependent on the relative character and strength of demand in the local labour market. The worry must be that in areas of weakest demand, where the majority of vacancies are low paid and insecure, StepUP jobs will be more attractive than unsubsidised jobs and participants will be reluctant to move out of the transitional sector, thus producing a type of lock-in. For instance, Bivand et al.'s recent report on StepUP suggests that 58 per cent of the unsubsidised jobs in the Pilot areas are either fixed-term, temporary, seasonal, casual or part-time. The report comments that clients may well prefer StepUP jobs to the opportunities in the normal labour market:

> A StepUP job is a more certain contract than seasonal work, casual or agency temping, and offers higher rewards than part-time work for the same hourly rate. Job quality can be described in terms of both income and certainty. In these terms StepUP jobs are both certain and give a higher weekly (though not necessarily hourly) income than jobs taken by most control areas respondents and jobs in unsubsidised areas (Bivand et al., 2004, p. 61).

While StepUP is a welcome development, which should not be solely evaluated in terms of its immediate job attainment targets, its success will be strongly influenced by the capability of the local labour markets to absorb those completing the contracts. There seems little doubt that even new TEPs in depressed local labour markets will be insufficient to compensate for endemically weak employment demand. But, as we have argued, there are nevertheless strong equity grounds for introducing a universal guarantee of a job of this sort, and indeed a longer contract, after a year's unemployment.

The unavoidable implication is that New Deal programmes in difficult local labour markets need to be supplemented by measures to address the demand for labour. It is impossible to significantly improve the quality of the contract offered to New Deal clients without reforming local labour markets. As Handler (2004) has recently argued, there is no point in having active labour market policies if good jobs are not available:

> There cannot be a separation between welfare policy and the economy. There are many obvious reasons why there have to be jobs for workfare clients. One that I would like to emphasise is that strong demand will reduce the pressures on the government agencies by facilitating the normal incentives of the clients (p. 269).

Where demand is strong, welfare agencies can concentrate on the more difficult cases. But there has been little attention given to how better to integrate the New Deal with employment creation policies. For instance, the myriad of urban regeneration initiatives tried in recent years have been poorly integrated with New Deal programmes and they appear to have originated and to have been run from separate policy silos. In general, despite the plethora of community economic development schemes (CEDs) and area-based regeneration initiatives (ABIs) (there are up to 27 different initiatives working in any English region), the record on achieving local regeneration has been disappointing (Gripaios, 2002; see also Armstrong et al., 2002; Armstrong, 2004). Most initiatives have been single issue, top-down and short-term schemes focused on improvement in very small areas and it is to be hoped that, in future, both Regional Development Agencies and Local Strategic Partnerships will be able to improve coordination and work towards more locally decentralised and holistic approaches (ODPM: Housing, Planning, Local Government and the Regions Committee, 2003). In 2004, however, Job Centre Plus was found to be a core member of only 37 per cent of existing Local Strategic Partnerships, which is a clear indication of institutional fragmentation and the lack of priority attached to local employment issues (SEU, 2004). To date there is little evidence that CEDs and ABIs have had a long-term impact on local labour markets. As Brennan et al. (2000) argue, the need for new initiatives to tackle geographical concentrations of social exclusion by targeting resources on those in need and improving the physical and environmental characteristics of the areas concerned, has probably never been greater. Similar views have also been expressed in the US context (see e.g. Freeman and Gottschalk, 1998).

The merit and scale of targeted job creation policies has been keenly debated (see Houston, 1998). Webster (1999), for example, argues that locally targeted job creation is the crucial missing element in current labour market and social inclusion policy. He suggests that more investment needs to be made in updating the physical and transport infrastructure in older industrial cities in order to support manufacturing employment. He argues that the property needs of manufacturing firms could be met by developing brownfield land and that this should be accompanied by strategic road and rail investment, as well as discriminatory corporate taxation. He also argues (Webster, 2000) that the distance decay in commuting means that unless new jobs are located within about 3 miles of the target unemployment blackspots, their residents will not have access to any significant share of them.

On the other hand, Gordon (1999, 2001, 2003), drawing on research into local labour markets in London, argues that this approach may be less effective than its protagonists claim, as local labour markets can be highly permeable ('leaky'), in the sense that many of the jobs created in high unemployment localities may be taken by individuals from adjoining or nearby local labour markets, so that many of the net new vacancies created end up in surrounding areas, rather than in the area where they were intended. Research in the US has found that high proportions of new jobs in local areas are eventually absorbed by net migration (Bartik, 2001). Thus, according to Gordon (2003):

> The disadvantages of locally targeted job creation are threefold. Firstly, that associated costs will often be significantly higher; secondly that the localised

focus encourages a gross under-estimate of the scale of job creation required to lower unemployment substantially, given the high level of leakage. And thirdly, that it distracts attention from the forms of intervention required to make disadvantaged residents effective competitors for jobs accruing inside or outside the area (p. 87).

The implication of Gordon's argument is that job creation employment measures need to target a less 'leaky' geographical labour market area, such as the wider metropolitan or regional scale. In his words:

> On the demand side, the pressure of demand for labour experienced across the broader region within which areas of concentrated unemployment are located matters more. Within these regions, a high tide of demand raises all ships, to the particular benefit of groups of the least advantaged people and areas (1999, p. 201).

This view suggests, then, that while the New Deal seeks to tackle the unemployment problem in terms of local supply-side measures, stimulating the demand side – expanding the number and range of jobs – is best tackled at a larger, regional scale. But how feasible is this? And is it the correct approach?

Traditionally, UK regional policy was very much focused on equalising employment opportunities across the country, and numerous measures were used to redistribute investment and jobs from the relatively prosperous South-East and West Midlands regions to the relatively depressed northern areas of the country. Various claims have been made for the job creation success of post-war regional policy: for example, that up to the late-1970s it had helped to create as many as 500,000 jobs in the higher unemployment (or 'assisted') regions. Others contest these estimates, or at least point to the fact that many of the jobs proved to be vulnerable when the winds of economic slow-down and restructuring began to blow in the 1970s, and especially the 1980s. The old regional policy fell into disrepute as regional unemployment disparities widened. Since then, partly because it came to be perceived as increasingly ineffective, partly because of the shifting economic ideologies of the national policy paradigm under Thatcher and, more recently, under New Labour, regional policy has undergone a major reconfiguration. Under New Labour, especially, a new regional policy model has emerged in which the primary focus is no longer on equalising demand and employment opportunities across the country's regions, but one of raising the competitiveness, productivity and innovativeness of each and every region, and thence the nation as a whole (Balls and Healy, 2000; HM Treasury and DTI, 2001; Adams, Robinson and Vigor, 2003; HM Treasury, 2003). In terms of reducing regional economic disparities across the nations, the main objective is to equalise regional growth rates. Equalising growth rates does nothing, of course, to narrow differences in per capita GDP levels across the regions: indeed, regional differences in per capita GDP will in fact continue to widen. And in any case it is not clear how equalising growth rates is supposed to translate into employment growth: there is no explicit aim to equalise employment opportunities across regions and localities. Critics argue that this new policy model will do little to solve the severe jobs gaps that exist in northern regions and cities (Anyadike-Danes et al., 2000; Fothergill, 2005).

To compound the debate, is the regional level actually the best scale at which to seek to stimulate the demand for labour? The fact is that a high level of demand in a region may be a necessary, but is not a sufficient, condition for improving job opportunities in each and every locality in the region in question. It is well known that regional and local disparities tend to narrow in times of high national economic demand: a rising tide does tend to lift most boats. But it does not necessarily raise all boats equally: some tend to respond much more than others. Much depends on the condition of the boats! Economically weak regions and localities may be much less responsive to general increases in national demand than are economically dynamic regions and localities. Thus local deficiencies in demand for labour can persist even during conditions of national economic boom. And, as we saw in Chapter 2, persistent deficiencies in the local demand for labour can in turn set in train a host of supply-side processes that merely serve to reinforce the local unemployment problem (Figure 2.10). The importance of migration and consequence leakage effects is more of an argument for carefully targeting demand-side policies on the most disadvantaged areas and people, than it is a justification for abandoning demand-side policies.

Gordon's argument that region-wide demand conditions are more important than locally targeted job creation measures is thus not unproblematic: there is no guarantee that a high level of regional demand will boost employment opportunities in each and every locality. Indeed, the existence of local pockets of persistent high unemployment in the southeast, the most dynamic and prosperous region in the UK, testifies to this. Notwithstanding the possibility of 'leakage effects' of the sort discussed by Gordon, job creation measures may need a specifically local dimension if they are to work in conjunction with local supply-side orientated New Deal programmes. In fact, given that pockets of high unemployment are typically also areas with poor housing, rundown physical environments and inferior infrastructures, job creation, neighbourhood improvement, and economic regeneration all ought to go together.

In other words, the current emphasis on what might be termed a *micro supply-side approach* to the unemployment problem needs to be accompanied by corresponding targeted or *micro-demand-side policies*. A mix of micro-demand-side policies is required, given that measures to upgrade the housing, environmental and business conditions of job-poor, high-unemployment localities may take some time to have effect. Thus, at the same time, in order to have a more immediate impact, other micro-demand-side policies are necessary to try to stimulate job creation among local employers. Wage subsidies to employers to hire disadvantaged workers can succeed in raising demand for labour for those workers, even if only modestly (Freeman and Gottschalk, 1998). Targeting subsidies in high-unemployment local labour markets should help to minimise any inflationary effects, and there is evidence that increases in demand for disadvantaged groups can increase their soft skills and reputation with employers and hence their long-term employment and earnings (Bartik, 2001). What is important is that the subsidy is attached to jobs, not to individuals, so there is none of the stigma associated with schemes in which individuals of a given type 'carry' subsidies with them to employers. At the same time, in order to reduce the leakage effects caused by in-migration, the jobs created should be linked to disadvantaged individuals through the action of local intermediaries (Bartik, 2001). It has also to be recognised that such subsidised jobs may reduce the incentive to acquire training. This effect, however, could be offset by training subsidies and customized training programmes designed in co-operation with local employers. Furthermore, the scale of such subsidies could vary according local labour market conditions, so that an element of local targeting is included. In addition much more attention should be paid to the possibility of directing public service employment growth towards the most depressed localities. New Deal does, of course, include subsidised employment and training options but these have not been effective micro-demand-side instruments, and could be further developed and integrated, as well as made locally varying in scale, rather than nationally uniform.

What is clear is that the dynamics of employment growth, recruitment, job queues and vacancy chains within (and between) local labour markets need more research. But it is important that these disagreements on the scale of policy intervention are not used to justify inaction on job creation. The search for a single scale for intervention is surely a false trail, as job creation policies will work best when simultaneously pursued at local, metropolitan regional and national scales.

## A Final Note on Geography and Workfare

To conclude, we have argued that the New Deal programmes appear to be a fragile mixture of workfare with some human-capital and individual support measures. The future of this combination is not predetermined but may well be reaching an important branching or decision point. On the one hand, the Government is promising to reform the New Deal programmes in order to provide greater decentralisation and local decision-making in an effort better to respond to local problems and to introduce targeted measures designed to address the particular difficulties of areas of entrenched labour market distress. On the other hand, the work-first supply-side agenda appears to be increasingly dominant and unquestioned, as shown by ever more grand claims about its effectiveness and the universal availability of jobs and vacancies. In our view these two agendas are not compatible and one will have to give. As we have argued throughout, a precondition to responding to local labour market problems is a proper recognition of the consequences of local failures in employment demand as well as in labour supply. The abstract thinking that emphasises the benefits of labour market participation, and the civic value of imagined contracts between individuals and the state, needs to appreciate that when actually implemented and grounded in different local economic contexts, workfare and active labour market policies are bound to have very different practical consequences, depending on the extent and nature of local labour market opportunities. Without this recognition, the policy improvements outlined here will amount to little more than elaborate window-dressing, and the policy agenda will remain more one of 'centralised localism' rather than of real local decentralisation, and thus be unable to respond properly to local problems. We suspect, however, that there are distinct limits to how far the UK Government – of whatever persuasion – is willing to devolve workfare policy design and responsibility down to the local level.

And while the New Deal has rightly highlighted the importance of the local dimensions of labour supply, it thus far seems to have done little to increase the willingness of policy-makers to recognise the quintessentially local nature of labour demand, how the characteristics of local labour supply are in part shaped by those of local labour demand, and thus the need also to intervene locally on this side of the labour market. As we have argued throughout this book, geography enters on both sides of the labour market 'equation', not just as a locally varying context for implementing nationally designed micro-supply-side policies, but also as a locally varying context in which the job creation and the employment process are in large part actually constituted. Geography thus complicates workfare-type policies in two key senses: by shaping the underlying unemployment problem such policies are meant to address, and by conditioning the success with which they help solve it. For both reasons, consideration of place is indispensable for assessing whether workfare works.

# Notes

## Chapter 1

1 Similarly, Jessop (1994, 1999) argued that the national Keynesian welfare state is being superseded by a post-Fordist Schumpeterian workfare state whose primary goal is to ensure continuous economic flexibility and innovation by intervening primarily on the supply-side.

2 In effect this is a rediscovery or reworking of the so-called Okun negative (downward-sloping) trade-off between economic efficiency and social equity (Okun, 1975) whereby the pursuit of the latter compromises the former.

3 Giddens (2000) accepts that some degree of economic redistribution is necessary to provide equality of opportunity as, in its absence, unequal outcomes are transmitted across generations, thereby denying children equal opportunities. However, his comments on redistribution are vague and sit rather uneasily with his claims that income transfers and high taxation are outdated and counterproductive. This idea of a 'social investment state' is not new; the phrase was in fact used by Keynes in the 1930s, in making the case for state intervention in the national economy, though he tended to emphasise investment in physical infrastructure and public services rather than in human capital.

4 Sanctions may be applied to those who refuse opportunities (such as refusing to attend an interview or going absent from a placement) and may involve the loss of two or four weeks benefit. The sanctions imposed in each year between 1999 and 2000 and 2003 and 2004 have varied between 3.4 and 4.7 percent of total starts on the NDYP (House of Commons, 2004)

5 This was the major part of the total budget for all the New Deal programmes of £5.2 billion. Some have criticised the balance of funding for giving too much priority to youth unemployment, at the expense of spending on the New Deals for older age groups (Alcock et al., 2003).

# Chapter 3

1  These claims are partly based on the findings of national macroeconomic evaluations which suggested that the programme had been cost effective and a modest success. In 2001–2 they concluded that it had cut youth unemployment by between 35,000 and 40,000 and that unemployed young men were about twenty percent more likely to get a job as a result of the policy, thereby increasing youth employment by over 17,000 (White and Riley, 2002; Van Reenen, 2001).

2  This reflects the government's tendency to argue that social problems are localised and that partnership activity and small area policy zones can make a real difference (Mohan 2000; Smith, 1999; Chatterton and Bradley, 2000; Foley and Martin, 2000). This stance has been heavily influenced by the view that long term unemployment is a product of micro-level supply-side failures which mean that individuals in certain local areas become unemployable and unable to benefit from local employment growth (Campbell et al., 1998). Thus mainstream policies should be 'bent' towards poorer localities and made more locally responsive (Social Exclusion Unit, 1998).

3  It is not possible to calculate youth unemployment rates for all 144 UoDs. We have calculated these rates by amalgamating estimates of 18–24 populations by local authority districts (based on the Labour Force Survey, Local Authority Database for 1997 and 1998) into UoD areas. For a number of local authority districts these estimates are too small to be reliable, and hence certain UoDs had to be omitted from this exercise.

4  The 'unknown destinations' measure counted those participants who left the programme but who did not tell the Employment Service the reason for this. It included those leaving to take up jobs, as well as those who left the registered labour force.

5  This updates and corrects the claim in the earlier NIESR report for the Employment Service that 'Initial analysis of inflows to unemployment does not suggest that the NDYP is just circulating young people through the programme and back on to the claimant count'(Anderton et al., 1999, p. 12).

6  The local authority database provides estimates of 18–24 populations by local authority district but if these are less than 6,000 which is the case for many rural districts they are considered statistically unreliable (see note 2).

7  A fuller discussion of how unemployment claimant flows relate to participation rates is given in Martin and Sunley (1999); see also Beatty et al. (2002).

8  Again, we are using the Labour Force Survey, local authority database for 1997 and 1998 estimates of 18–24 populations to generate an estimate of the source population. Unfortunately, after 1998 the series was discontinued.

9  The figures for Cambridge and Edinburgh and East Lothian Units of Delivery in particular should be read as estimates only as they include outlying rural LADs with small total youth populations.

10  This is in distinction to the Treasury's (2000) use of the claimant count as the basis for its assertion that there is little evidence of a real jobs gaps.

11    If those on self-employment support are included, the figure increases to nearly 31 per cent (Bivand, 2003).

12    Over a quarter of participants are recorded as having a form of disability (Wilkinson, 2003).

## Chapter 4

1    For instance, it is noticeable that the rates of clients leaving to 'unknown destinations' is very high in London (between 30 to 40 per cent of all leavers). The Department for Work and Pensions argues that the actual destinations of this group mirror the destinations of all leavers, but the degree to which this may vary at a local level is unknown.

2    For example, the tendency to sanction New Deal participants for not accepting options has varied enormously between different regions of Britain, with no obvious explanation for the varying sanction practices apart from management differences (see Centre for Economic and Social Inclusion *Working Brief*, 121, February 2001).

## Chapter 5

1    Namely routine unskilled; operative/assembly; sales; personal/protective services; craft skilled; clerical/secretarial; associate professionals/technicians; professionals; managers/administrators.

2    The initial rate of £3.60 for those aged under 21 was raised to £3.70 in October 1999. From October 2003 the adult rate was raised to £4.50 an hour and the development (18–21 year old) rate to £3.80 per hour. A new rate of £3 per hour for 16–17 year olds was introduced in October 2004 when the 18–21 rate also increased to £4.10 per hour.

3    This requirement seems to have been brought late into the running of NDYP. Employers who had participated in the early stages (i.e. in 1998 and 1999) did not report having had to return the individual training plan to start being paid the subsidy. One must take into account that the fieldwork in Edinburgh was conducted in December 2000 / January 2001 and that this case study was not on a level playing field with other UoDs in which the fieldwork had been conducted months beforehand, which leaves time for policy adjustments.

## Chapter 6

1    Eleven Employer Coalitions have been established, mostly in inner city areas. They are made up of a dozen or so private sector employers, who volunteer their involvement, together with senior public sector partners. They are usually chaired by a prominent local business leader.

2   The Adviser Discretion Fund was introduced in July 2001 and allowed personal advisers in Employment Zones to make awards of up to £300 to help their clients purchase clothing, travel passes and with childcare.

# Chapter 7

1   Marshall and MacFarlane (2000) suggested that the average cost of ILMs in Britain was £13,860 per place per year, but Finn and Simmond's (2003) survey points to a figure of £11,134 for all ILMs in 2002/3, and a figure of £10,638 for New Deal based ILMs.
2   The areas are East Ayrshire, Dundee, Sunderland, Knowsley, Burnley, Manchester, Oldham, Barnsley and Rotherham, Bradford, Leeds, Sheffield, Cardiff, Wrexham, Coventry, Sandwell, Great Yarmouth, Hackney, Lambeth, Greenwich and Bristol.

# Bibliography

Adams, J., Robinson, P. and Vigor, A. (2003) *A New Regional Policy for the UK*, London: Institute for Public Policy Research.

Alcock, P., Beatty, C., Fothergill, S., Macmillan, R. and Yeandle, S. (2003) *Work to Welfare: How Men Become Detached from the Labour Market*, Cambridge: Cambridge University Press.

Allard, S. (2002) The Urban Geography of Welfare Reform: Spatial Patterns of Caseload Dynamics in Detroit, *Social Science Quarterly 83*, 4, pp. 1044–1062.

Allard, S., Tolman, R. and Rosen, D. (2003) The Geography of Need: Spatial Distribution of Barriers to Employment in Metropolitan Detroit, *Policy Studies Journal 31*, 3, pp. 293–307.

Allen J. and Henry, N. (1997) Ulrich Beck's Risk Society at Work: Labour and Employment in the Contract Service Industries, *Transactions of Institute of British Geographers 22*, pp. 180–196.

Anderton, B., Riley, R. and Young, G. (1999) *The New Deal for Young People: First Year Analysis of Implications for the Macroeconomy*, Employment Service Research and Development Report/NIESR, ESR33.

Annesley, C. (2003) Americanised and Europeanised: UK Social Policy since 1997, *British Journal of Politics and International Relations 5*, 2, pp. 143–165.

Anyadike-Danes, M., Fothergill, S., Glyn, A., Grieve-Smith, J., Kitson, M., Martin, R.L., Rowthorn, R., Turok, I., Tyler, P. and Webster, D. (2000) *Tackling the Regional Jobs Gap*, London: Employment Policy Institute.

Armstrong, H. (2004) The Role of Community Economic Development Policies in Regional Policy, in I. Hara and S. Tokunaga (Eds.), *The Community Economic Development Model: Prospects for Regional Development Models*, Tokyo: Bunshindou, pp. 1–31.

Armstrong, H., Kehrer, B., Wells, P. and Wood, A. (2002) The Evaluation of Community Economic Development Initiatives, *Urban Studies 39*, 3, pp. 457–481.

Atkinson, J. (1999) *The New Deal for Young Unemployed People: A Summary of Progress*, London: Employment Service Research and Development Report ESR13.

Atkinson, J., Kodz, J., Dewson, S. and Eccles, J. (2000) *Evaluation of New Deal 50Plus: Qualitative Evidence from Clients, Employment Service Report 52*, London: Department for Education and Employment.

Auer, P. (2000) *Employment Revival in Europe: Labour Market Success in Austria, Denmark, Ireland and the Netherlands*, Geneva: ILO.

Baddeley, M., Martin R. and Tyler, P. (2000) Regional Wage Rigidity in the European Union and the United States Compared, *Journal of Regional Science 40*, 1, pp. 113–140.

Balls, E. and Healey, J. (Eds.) (2000) *Towards a New Regional Policy*, London: Smith Institute.

Bank of Scotland and Scottish Enterprise Edinburgh and Lothian (2000) *The Financial Service Sector Partnership for New Deal, Evaluation Report*, November 2000, Edinburgh.

Barbier, J.-C. (2001) *Welfare to Work Policies in Europe: The Current Challenges of Activation Policies*, Paris: Centre d'etudes de l'emploi: Document de travail, 11.

Barbier, J.-C. and Ludwig-Mayerhofer, W. (2004) Introduction: The Many Worlds of Activation, *European Societies 6*, 4, pp. 423–436.

Bartik, T. (2001) *Jobs for the Poor: Can Labor Demand Policies Help?* New York: Russell Sage Foundation.

Bean, C. (1994) European Unemployment: A Survey, *Journal of Economic Literature 32*, pp. 573–619.

Beatty, C. and Fothergill, S. (2003) The Detached Male Workforce, in P. Alcock, C. Beatty, S. Fothergill, R. Macmillan and S. Yeandle, *Work to Welfare: How Men Become Detached from the Labour Market*, Cambridge: Cambridge University Press, pp. 79–110.

Beatty, C., Fothergill, S., Gore, T. and Herrington, A. (1997) *The Real Level of Unemployment*, CRESR, Sheffield: Sheffield Hallam University.

Beatty, C., Fothergill, S. and Gore, T. and Green, A. (2002) The Real Level of Unemployment 2002, CRESR, Sheffield: Sheffield Hallam University.

Beck, U. (1992) *Risk Society: Towards a New Modernity*, London: Sage.

Beland, D., Vergniolle, F. and Waddan, A. (2002) Third Way Social Policy: Clinton's Legacy? *Policy and Politics 30*, 1, pp. 19–30.

Benito, A. and Oswald, A. (1999) Commuting in Great Britain during the 1990s, http://www.andrewoswald.com.

Benner, C. (2002) *Work in the New Economy: Flexible Labour Markets in Silicon Valley*, Oxford: Blackwell.

Bennett, R. and Payne, D. (2000) *Local and Regional Economic Development: Re-negotiating Power under Labour*, Aldershot: Ashgate.

Berthoud, R. (2003) *Multiple Disadvantage in Employment*, York: Joseph Rowntree Foundation.

Bivand, P. (2001) New Deal Sanctions, *Working Brief 121*, February 2001, available at http://www.cesi.org.uk.

Bivand, P. (2003) Lone Parent New Deal Meetings Decline in Effectiveness, *Working Brief 151*, pp. 19–21.

Bivand, P., Britton, L., Morrin, M. and Simmonds, D. (2004) *Evaluation of StepUP: Interim Report*, London: DWP 186.

Blackmore, M. (2001) Mind the Gap: Exploring the Implementation Deficit in the Administration of the Stricter Benefits Regime, *Social Policy and Administration* 35, 2, pp. 145–162.

Blair, T. (1999a) Should the Welfare State Be Reformed? – The Case for, *Observer*, Sunday 23 May.

Blair, T. (1999b) Speech on New Deal, 22 June 1999, http://www.number-10.gov.uk.

Blank, R. (2000) Fighting Poverty: Lessons from Recent U.S. History, Distinguished Lecture on Economics in Government, *Journal of Economic Perspectives* 14, 2, pp. 3–19.

Blank R. and Card, D. (2001) (Eds.) *Finding Jobs: Work and Welfare Reform*, New York: Russell Sage Foundation.

Blank, R., Card, D., and Robins, P. (2001) Financial Incentives for Increasing Work and Income among Low-Income Families, in D. Card and R. Blank (Eds.), *Finding Jobs: Work and Welfare Reform*, New York: Russell Sage Foundation, pp. 373–419.

Blundell, R. (2001) Welfare-to-Work: Which Policies Work and Why? Keynes Lecture in Economics 2001, available at http://www.ifs.org.uk/ conferences/keynes2001.pdf.

Blundell, R., Reed H., Van Reenen, J. and Shephard, A. (2003) The Impact of the New Deal for Young People on the Labour Market: A Four Year Assessment, in R. Dickens, P. Gregg and J. Wadsworth (Eds.), *The Labour Market under New Labour: The State of Working Britain 2003*, Basingstoke: Palgrave Macmillan, pp. 17–31.

Bonjour, D., Dorsett, R. Knight, G., Lissenburgh, S., Mukerjee, A., Payne, J., Range, M., Urwin, P. and White, M. (2001) *New Deal for Young People: National Survey of Participants Stage 2*, London: DWP ESR67.

Branosky, N. (2004) Welfare to Work: Looking to the Future, *Working Brief 152*, pp. 17–19.

Brennan, A., Rhodes, J. and Tyler, P. (2000) The Nature of Local Area Social Exclusion in England and the Role of the Labour Market, *Oxford Review of Economic Policy 16*, 1, pp. 129–146.

Brodkin, E. and Kaufman, A. (1998) *Experimenting with Welfare Reform: The Political Boundaries of Policy Analysis*, Northwestern University, University of Chicago, Joint Center Poverty Research Working Paper, No. 1.

Brown, G. (2001) *Enterprise and the Regions*, Speech at UMIST on Monday 29 January 2001, www.hm–tresury.gov.uk/press/2001.

Brown, D., Dickens, R., Gregg, P., Machin, S. and Manning, A. (2001) *Everything under a Fiver: Recruitment and Retention in Lower Paying Labour Markets*, York: Joseph Rowntree Foundation.

Browne, D. (2004a) *Minister Pledges to Tackle Culture of Worklessness*, DWP Press Release, 1 April.

Browne, D. (2004b) Working Brief – Interview, Http://www.newdeal.gov.uk/ catalyst/data/13-2004/data/working-brief2.html.

Bryson, A., Knight, G. and White, M. (2000) *New Deal for Young People: National Survey of Participants: Stage 1*, Sheffield: Employment Service Research and Development Report ESR44.

Budd, A., Levine, P. and P. Smith (1988) Unemployment, Vacancies and the Long-Term Unemployed, *Economic Journal 98*, pp. 1071–1091.

Burden, T., Cooper, C. and Petrie, S. (2000) *'Modernising' Social Policy: Unravelling New Labour's Welfare Reforms*, Aldershot: Ashgate.

Burgess, S. (1989) How Does Unemployment Change? Mimeo, Department of Economics, University of Bristol, Bristol.

Burghes, L. (1992) *Working for Benefits: Lessons from America*, London: Low Pay Unit.

Burkitt, N. and Hughes, A. (2004) Tackling Local Pockets of Worklessness, http://www.socialexclusionunit.gov.uk.

Calmfors, L. (1994) *Active Labour Market Policy and Unemployment: A Framework for the Analysis of Crucial Design Features*, Paris: OECD Working Papers 40.

Calmfors, L. (1995) 'What can we expect from Active Labour Market Policy?', *Applied Economics Quarterly 43*, pp. 11–30.

Calmfors L. and Skedinger, P. (1995) Does Active Labour Market Policy Increase Employment? Theoretical Considerations and Some Empirical Evidence from Sweden, *Oxford Review of Economic Policy 11*, 1, pp. 91–109.

Campbell, M. (2000) Reconnecting the Long Term Unemployed to Labour Market Opportunity: The Case for a 'Local Active Labour Market Policy', *Regional Studies 34.7*, pp. 655–668.

Campbell, M., Sanderson, I and Walton, F. (1998) *Local Responses to Long Term Unemployment*, York: Joseph Rowntree Foundation.

Campbell, M, Foy, S. and Hutchinson, J. (1999) Welfare to Work: Lessons from the United Kingdom, in OECD (1999), *The Local Dimension of Welfare to Work: An International Survey*, Paris: OECD, pp. 197–222.

Capano, G. (2003) Administrative Traditions and Policy Change: When Policy Paradigms matter. The case of Italian Administrative Reform during the 1990s, *Public Administration 81*, 4, pp. 781–801.

Casebourne, J. (2003) *Work, Poverty and Welfare Reform: Welfare-to-work Programmes for Lone Parents in Depressed Local Labour Markets*, London: CESI.

Centre for Economic and Social Inclusion (2001) *Working Brief 121*, London: CESI.

Cerny, P. and Evans, M. (2004) Globalisation and Public Policy under New Labour, *Policy Studies 25*, 1, pp. 51–65.

Chatterton, P. and Bradley, D. (2000) Bringing Britain together? The Limitations of Area-Based Regeneration Policies in Addressing Deprivation, *Local Economy 15* 2, pp. 98–111.

Clarke, J. (2004) Dissolving the Public Realm? The Logics and Limits of Neo-liberalism, *Journal of Social Policy 33*, 1, pp. 27–48.

Clarke, J. and Newman, J. (1997) *The Managerial State: Power, Politics and Ideology in the Remaking of Social Welfare*, London: Sage.

Clarke, J., Gewirtz, S. and McLaughlin, E. (2000) *New Managerialism, New Welfare?* London: Sage.

Clarke, J., Langan, M. and Williams, F. (2001) Remaking Welfare: The British Welfare Regime in the 1980s and 1990s, in A. Cochrane, J. Clarke, and S. Gewirtz (Eds.), *Comparing Welfare States* (2nd edition), London: Sage, pp. 71–112.

Clasen, J. (2002) Unemployment and Unemployment Policy in the UK: Increasing Employability and Redefining Citizenship, Chapter 3 in J. Andersen, J. Clasen, W. Van Oorschot and K. Halvorsen (Eds.), *Europe's New State of Welfare: Unemployment, Employment Policies and Citizenship*, Bristol: Policy Press.

Clasen, J. (2003) Towards a New Welfare State or Reverting to Type? Some Major Trends in British Social Policy since the Early 1980s, *European Legacy 8*, 5, pp. 573–586.

Clasen, J. and Clegg, D. (2003) Unemployment Protection and Labour Market Reform in France and Great Britain in the 1990s: Solidarity versus Activation? *Journal of Social Policy 32*, 3, pp. 361–381.

Clasen, J., Gould, A. and Vincent, J. (1998) *Voices Within and Without: Responses to Long-Term Unemployment in Germany, Sweden and Britain*, Bristol: Policy Press.

Cochrane, A. (2004) Modernisation, Managerialism and the Culture Wars: Reshaping the Local Welfare State in England, *Local Government Studies 30*, 4, pp. 481–496.

Cochrane, A., Clarke, J. and Gewirtz, S. (2001)(Eds.) *Comparing Welfare States* (2nd Edition), London: Sage.

Considine, M. (2002) The End of the Line? Accountable Governance in the Age of Networks, Partnerships and Joined Up Services, *Governance: An International Journal of Policy, Administration, and Institutions 15*, 1, pp. 21–40.

Considine, M. (2003) Local Partnerships: Different Histories, Common Challenges – A Synthesis, Chapter 16 in OECD, *Managing Decentralisation: A New Role for Labour Market Policy*, Paris: OECD, pp. 253–272.

Cox, R. (1998a) The Consequences of Welfare Reform: How Conceptions of Social Rights are Changing, *Journal of Social Policy 27*, 1, pp. 1–16.

Cox, R. (1998b) From Safety Net to Trampoline, *Governance 11*, 4, pp. 397–414.

Coyle, D. (2001) *Paradoxes of Prosperity: Why the New Capitalism Benefits All*, London: Texere.

Cutler, T. and Waine, B. (2000) Managerialism Reformed? New Labour and Public Sector Management, *Social Policy 34*, 3, pp. 318–332.

Daguerre, A. (2004) Importing Workfare: Policy Transfer of Social and Labour Market Policies from the USA to Britain under New Labour, *Social Policy and Administration 38*, 1, pp. 41–56

Daguerre, A. and Taylor Gooby, P. (2004) Neglecting Europe: Explaining the Predominance of American ideas in New Labour's Welfare Policies Since 1997, *Journal of European Social Policy 14*, 1, pp. 25–39.

Danziger, S.K. and Danziger, S. (1995) Will Welfare Recipients Find Work when Welfare Ends?, in Sawhill, S. (Ed.), *Welfare Reform: An Analysis of the Issues*, Washington, D.C.: Urban Institute, pp. 41–44.

Davies, V. and Irving, P. (2000) *NDYP: Intensive Gateway Trailblazers*, Sheffield: ES Research and Development Report ESR50.

Deacon, A. (2002) *Perspectives on Welfare: Ideas, Ideologies and Welfare Debates*, Buckingham: Open University Press.

Dean, H. (2001) Welfare Rights and the 'Workfare' State, *Benefits, January–February*, pp 1–4.

Department for Education and Employment (DfEE) (1998) *Design of the New Deal for 18–24 Year Olds*, London: HMSO.

Department for Education and Employment (2000) New Deal Close to Paying for Itself – Jowell, Press Release 240/00, 25 May.

Department for Education and Employment (2001a) 'Golden Hellos' for Employers at Heart of New, Simpler Employment Drive, Press Notice 2001/0021, 17 January.

Department for Education and Employment (2001b) New Deal Success Undercounted – Survey, Press Notice 2001/0053, 31 January.

Department for Education and Employment, Department for Social Security, HM Treasury (2001) *Towards Full Employment in a Modern Society*, Cm 5084, London: TSO.

Department for Social Security (DSS) (1998) *A New Contract for Welfare: The Gateway to Work*, London: DSS.

Department for Work and Pensions (DWP) and HM Treasury (2002) *UK Employment Action Plan, 2002*, London: DWP.

Department for Work and Pensions (2002) Stepping Stones to Jobs for Long Term Unemployed, Press Release 8 May.

Department for Work and Pensions (2004a) *Building on New Deal: Local Solutions meeting individual needs*, http://www.dwp.gov. uk/publications/ dwp/2004/buildingonnewdeal.

Department for Work and Pensions (2004b) Statistics on New Deal Immediate Destinations, http://www.dwp.gov.uk/asd/ndyp.asp.

Dickens, R. and Ellwood, D.T. (2000) Whither poverty in Great Britain and the United States? The Determinants of Changing Poverty and whether Work Will Work, *Discussion Paper 506*, Centre for Economic Performance, London.

Dickerson, A. (2003) *Exploring Local Areas, Skills and Unemployment: The Relationship between Vacancies and Local Unemployment*, London: DWP.

Disney, R. and Carruth, A. (1992) *Helping the Unemployed: Active Labour Market Policies in Britain and Germany*, London: An Anglo-German Foundation Report.

Dolowitz, D. (1998) *Learning from America: Policy Transfer and the Development of the British Workfare State*, Brighton: Sussex Academic Press.

Driver S. and Martell, L. (1998) *New Labour: Politics after Thatcherism*, Bristol: Policy Press.

Dunkerley, M. (1996) *The Jobless Economy? Computer Technology in the World of Work*, Cambridge: Polity Press.

Dworkin, R. (2000) *Sovereign Virtue: The Theory and Practice of Equality*, Harvard: Harvard University Press.

Easterlow D. and Smith, S. (2003) Health and Employment: Towards a New Deal, *Policy and Politics 31*, 4, pp. 511–533.

Education and Employment Committee (1998) *Second Report: The New Deal Volume 1 and Volume 2 Minutes of Evidence*, HoC 1997–98, HC 263–I & II.

Education and Employment Committee (2000) *New Deal for Young People Two Years On*, House of Commons 1999–2000, HC510.

Education and Employment Committee (2001a) *New Deal: An Evaluation*, Fifth Report, Proceedings, Minutes of Evidence and Appendices, HC 58, London: TSO.

Education and Employment Committee (2001b) Government's Response to the Fifth Report from the Committee Session 2000–01 New Deal: An Evaluation, HC 519, London: HMSO.

Elam, G. and Snape, D. (2000) *New Deal for Young People: Striking a Deal with Employers*, London: DWP RR36.

Elliot, L. and Atkinson, D. (1998) *The Age of Insecurity*, London: Verso.

Ellwood, D. (2000) Anti-Poverty Policy for Families in the Next Century: From Welfare to Work – and Worries, *Journal of Economic Perspectives 14*, 1, pp. 187–198.

Employment Service (2000) *New Deal: The Lessons So Far*, Sheffield: Employment Service.

Esping Andersen, G. (1990) *The Three Worlds of Welfare Capitalism*, Princeton: Princeton University Press.

Esping Andersen, G. (1996) *Welfare States in Transition: National Adaptations in Global Economies*, London: Sage.

Etherington, D. (1998) From Welfare to Work in Denmark: An Alternative to Free Market Policies? *Policy and Politics 26*, 2, pp. 147–161.

Etherington, D. and Jones, M. (2004) Whatever Happened to Local Government?: Local Labour Market Policy in the UK and Denmark, *Policy and Politics 32*, 2, pp. 137–50.

Etzioni, A. (1999) *The New Golden Rule: Community and Morality in a Democratic Society*, New York: Basic Books.

European Commission (2000) *Acting Locally for Employment: A Local Dimension for the European Employment Strategy*, Brussels: Commission of the European Communities, COM(2000) 196.

Evans, C., Nathan, M. and Simmonds, D. (1999) *Employability through Work*, CLES Research Paper No. 2, Manchester: Centre for Local Economic Strategies.

Evans M., McKnight, A. and Namazie, C. (2002) *New Deal for Lone Parents: First Synthesis Report of the National Evaluation*, London: DWP WAE116.

Evans, M., Eyre, J., Millar, J. and Sarre, S. (2003) *New Deal for Lone Parents: Second Synthesis Report of the National Evaluation*, London: DWP 163.

Ferguson, R. (2002) Rethinking Youth Transitions: Policy Transfer and New Exclusions in New Labour's New Deal, *Policy Studies 23*, 3–4, pp. 173–190.

Fevre, R. (2000) *The Sociology of the Labour Market*, London: Harvester Wheatsheaf.

Fieldhouse, E., Kalra, V. and Alam, S. (2002a) How New Is the New Deal? A Qualitative Study of the New Deal for Young People on Minority Ethnic Groups in Oldham, *Local Economy 17*, 1, pp. 50–64.

Fieldhouse, E., Kalra, V. and Alam, S. (2002b) A New Deal for Young People from Minority Ethnic Communities in the UK, *Journal of Ethnic and Migration Studies 28*, 3, pp. 499–513.

Finegold, D. (1999) Creating Self-Sustaining, High-Skill Ecosystems, *Oxford Review of Economic Policy 15*, 1, pp. 60–81.

Finegold, D. and Soskice, D. (1988) 'The failure of British Training: Analysis and Prescription', *Oxford Review of Economic Policy 4*, 3, pp. 21–53.

Finn, D. (1999) Job Guarantees for the Unemployed: Lessons from Australian Welfare Reform, *Journal of Social Policy 28*,1, pp. 53–71.

Finn, D. (2000) Welfare to Work: The Local Dimension, *Journal of European Social Policy 10*, 1, pp. 43–57.

Finn, D. (2003) The 'Employment-First' Welfare State: Lessons from the New Deal for Young People, *Social Policy and Administration 37*, 7, pp. 709–724.

Finn, D. and Simmonds, D. (2003) *ILMs in Britain and an International Review of Transitional Employment Programmes*, London: DWP Research Report 173.

Fletcher, D. R. (2004) Demand-Led Programmes: Challenging Labour Market Inequalities or Reinforcing them? *Environment and Planning C; Government and Policy 22*, pp. 115–128.

Foley, P. and Martin, S. (2000), A New Deal for the Community? Public Participation in the Regeneration and Local Service Delivery, *Policy & Politics 34*, 4, pp. 479–492.

Forslund, A. and Krueger, A. (1994) *An Evaluation of the Swedish Active Labour Market Policy: New and Received Wisdom*, Cambridge, Mass: NBER Working Paper 4802.

Forth, J. Metcalf, H. and Millward, N. (2002) London's Unemployment in the 1990s: Tests of Demand-Side Explanations for its Relative Growth, London: CEPR, DP 203.

Fothergill, S. (2005) A New Regional Policy for Britain, *Regional Studies, 39* (forthcoming).

Freedland, M. and King, D. (2003) Contractual Governance and Illiberal Contracts: Some Problems of Contractualism as an Instrument of Behaviour Management by Agencies of Government, *Cambridge Journal of Economics 27*, pp. 465–477.

Freeman, R. and Gottschalk, P. (Eds.) (1998) *Generating Jobs: How to Increase Demand for Less-Skilled Workers*, New York: Russell Sage Foundation.

Friedlander, D. and Burtless, G. (1995) *Five Years After: The Long-Term Effects of Welfare to Work Programs*, New York: Russell Sage.

Gardin, L. and Laville, J.-L. (2000) Les initiatives locales en Europe, *Travail et Emploi 81*, pp. 53–66.

Geddes, M. (2000) Tackling Social Exclusion in the European Union? The Limits to the New Orthodoxy of Local Partnership, *International Journal of Urban and Regional Research 24*, 4, pp. 782–800.

GHK, Cambridge Econometrics and Bannock Consulting (2004) *Understanding Performance Variation*, London: DWP Report 194.

Giddens, A. (1998) *The Third Way: The Renewal of Social Democracy*, Cambridge: Polity Press.

Giddens, A. (2000) *The Third Way and its Critics*, Cambridge: Polity Press.

Giddens, A. (2002) *Where Now for New Labour?* Cambridge: Polity Press.

Gilbert, N. (2002) *Transformation of the Welfare State: The Silent Surrender of Public Responsibility*, Oxford: Oxford University Press.

Glendinning,C., Powell, M. and Rummery, K. (2002) (Eds.) *Partnerships, New Labour and the Governance of Welfare*, Bristol: Policy Press.

Glennerster, H., Lupton, R., Noden, P. and Power, A. (1999) *Poverty, Social Exclusion and Neighbourhood: Studying the Area Bases of Social Exclusion*, CASE paper 22, London: CASE.

Glyn, A. and Erdem, E. (1999) *The UK Jobs Gap – Lack of Qualifications and the Regional Dimension*, Memorandum submitted to Employment SubCommittee Inquiry into Employability and Jobs.

Gordon, I.R. (1999) Targeting a Leaky Budget: The Case against Localised Employment Creation, *New Economy 6*, pp. 199–203.

Gordon, I.R. (2003) Unemployment and Spatial Labour Markets: Strong Adjustment and Persistent Concentration, in R.L. Martin and P.S. Morrison (Eds.), *Geographies of Labour Market Inequality*, London: Routledge, pp. 55–82.

Gough, I., Bradshaw, J., Ditch, J., Eardley, T. and Whiteford, P. (1997) Social Assistance in OECD countries, *Journal of European Social Policy 7, 1*, pp. 17–43.

Gouldner, A. (1960) The Norm of Reciprocity: A Preliminary Statement, *American Sociological Review 25*, pp. 161–178.

Granovetter, M. and Swedberg, R. (1995) (Eds.) *The Sociology of Economic Life*, Boulder: Westview.

Gray, A. (2001) *New Deal in Derbyshire*, Memorandum submitted to Education and Employment Committee's Fifth Report New Deal an Evaluation, published as Appendix 21, London: TSO, pp. 135–140.

Gray, A. (2002) European Perspectives on Welfare Reform: A Tale of Two Vicious Circles, *European Societies 4, 4*, pp. 359–380.

Gray, A. (2004) *Unsocial Europe: Social Protection or Flexploitation?* London: Pluto Press.

Green, A., Gregg, P. and Wadsworth, J. (1998) Regional Unemployment Change in Britain, in P. Lawless, R. Martin and S. Hardy (Eds.), *Unemployment and Social Exclusion*, London: Jessica Kingsley, pp. 69–93.

Green, A. and Owen, D. (1998) *Where Are the Jobless?* Bristol: JRF and Policy Press.

Green, A. and Owen, D. (2003) Skill Shortages: Local Perspectives from England, *Regional Studies 37, 2*, pp. 123–134.

Green, F. (1999) 'Training the workers', in P. Gregg and J. Wadsworth (Eds.), *The State of Working Britain*, Manchester: Manchester University Press, pp. 127–146.

Griffiths, R., Irving, P. and McKenna, K. (2003) *Synthesising the Evidence on Flexible Delivery*, London: DWP 171.

Griffiths R., Jones, G., Davies, V. and Youngs, R. (2003) *New Deal for Young People: Introducing a More Tailored Approach*, London: DWP 164.

Grimshaw, D., Ward, K.G., Rubery, J. and Beynon, H. (2001) Organisations and the Transformation of the Internal Labour Market, *Work Employment & Society 15, 1*, pp. 25–54.

Gripaios, P. (2002) The Failure of Regeneration Policy in Britain, *Regional Studies 34*, pp. 568–577.

Grover, C. and Stewart, J. (2000) Modernizing Social Security? Labour and its Welfare-to-Work Strategy, *Social Policy and Administration 34, 3*, pp. 235–252.

Gueron, J. and Pauly, E. (1991) *From Welfare to Work*, New York: Russell Sage.

Hales, J. and Collins, D. (1999) *New Deal for Young People: Leavers with Unknown Destinations*, Sheffield: Employment Service ESR21.

Hales, J., Collins, D., Hasluck, C. and Woodland, S. (2000) *New Deal for Young People and the Long-Term Unemployed – Survey of Employers*, London: DWP 58.

Hales, J., Taylor, R., Mandy, W. and Miller, M. (2003) *Evaluation of Employment Zones*, London: DWP

Hall, P. (1993) Policy Paradigms, Social Learning and the State: The Case of Economic Policy-making in Britain, *Comparative Politics 25, 3*, pp. 275–296.

Hall P. and Soskice, D. (2001) *Varieties of Capitalism: The Institutional Foundations of Comparative Advantage*, Oxford: Oxford University Press.

Hamblin, M. (1997) A Report on Workstart 3, Report for the Employment Service, Ref: 95C386/96C632, May.

Handler, J. (2004) *Social Citizenship and Workfare in the United States and Western Europe: The Paradox of Inclusion,* Cambridge: Cambridge University Press.

Hanson, S. and Pratt, G. (1992) Dynamic Dependencies: A Geographic Investigation of Local Labour Markets, *Economic Geography 68,* pp. 373–405.

Hasluck, C. (1999a) *Employers, Young People and the Unemployed: A Review of Research,* London: Employment Service Research and Development Report, ESR12.

Hasluck, C. (1999b) *Employers and the Employment Option of the New Deal for Young People: Employment Additionality and its Measurement,* London: Department for Education and Employment, Employment Service Research Report 14.

Hasluck, C. (2000) *New Deal for Lone Parents: A Review of Evaluation Evidence,* Employment Service Research Report 51, London: Department for Education and Employment.

Haughton, G., Jones, M., Peck, J., Tickell, A. and While, A. (2000) Labour Market Policy as Flexible Welfare: Prototype Employment Zones and the New Workfarism, *Regional Studies 34,* 7, pp. 669–680.

Hay, C. (1999) *The Political Economy of New Labour,* Manchester: Manchester University Press.

Hay, C. (2002) (Ed.) *British Politics Today,* Cambridge: Polity Press.

Helms, G. (2004) Municipal Policing Meets the New Deal, Working Paper, Dept of Geography and Geomatics, University of Glasgow, http://web.geog.gla.ac.uk/ online papers/ghelms001.pdf

Herzenberg, S., Alic, J. and Wial, H. (2000) *New Rules for a New Economy: Employment and Opportunity in Postindustrial America,* Ithaca: Cornell University Press.

Hillage, J. and Mitchell, H. (2003) *Evaluation of Employer Training Pilots,* Department for Education and Skills Report ETP1, London: DFES.

Hillage, J. and Pollard, E. (1998) *Employability: Developing a Framework for Policy Analysis,* DfEE Research Report 85, Nottingham: DfEE.

Hills, J., LeGrand, J. and Piachaud, D. (2001) *Understanding Social Exclusion,* Oxford: Oxford University Press.

HM Treasury (1997) *The Modernisation of Britain's Tax and Benefit System, Number 1, Employment Opportunity in a Changing Labour Market,* London: HM Treasury.

HM Treasury (2000) *The Goal of Full Employment: Employment Opportunity for All Throughout Britain,* London: HM Treasury.

HM Treasury (2003a) *A Full Employment Strategy for Europe,* London: HMSO.

HM Treasury (2003b) *A Modern Regional Policy for the United Kingdom,* London: HM Treasury.

HM Treasury and Department of Trade and Industry (2001) *Productivity in the UK: The Regional Dimension,* London: HM Treasury.

HM Treasury and DWP (2001a) *The Changing Welfare State: Employment Opportunity For All,* London: HMSO.

HM Treasury and DWP (2001b) *Productivity in the UK: 3 – The Regional Dimension,* London: HMSO.

HM Treasury and DWP (2003) *Full Employment in Every Region,* London: HMSO.

HM Treasury, Office of the Deputy Prime Minister and Department for Trade and Industry (2004) *Devolving Decision Making: Meeting the Regional Economic Challenge; Increasing Regional and Local Flexibility*, London: HMSO.

Holtham, G., Ingham, P. and Mayhew, K. (1998) New Deal May Not Be Just the Job, *Observer*, 22 February1998, Business, p. 4.

Hoogvelt, A. and France, A. (2000) New Deal: The Experience and Views of Clients in One Pathfinder City (Sheffield), *Local Economy 15*, 2, pp. 112–127.

Houston, D. (1998) Job Proximity and the Urban Employment Problem: Do Suitable Nearby Jobs Improve Neighbourhood Employment Rates? A Comment, *Urban Studies 35*, 12, pp. 2353–2357.

Huber, E. and Stephens, J. (2000) *Development and Crisis of the Welfare Sate: Parties and Politics in Global Markets*, Chicago: University of Chicago Press.

Hughes, M.A. (1996) Learning from the 'Milwaukee Challenge', *Journal of Social, Policy Analysis and Management 15*, 4, pp. 562–571.

Hyland, T. and Musson, D. (2001) Unpacking the New Deal for Young People: Promise and Problems, *Educational Studies 27*, 1, pp. 55–67.

Islam, F. (2001) Getting Ready to Spend: Interview with the Chancellor, *Observer*, 15 April, Business Section, p. 3.

Jackman R. and Savouri, S. (1999) Has Britain Solved the 'Regional Problem', in P. Gregg and J. Wadsworth (Eds.), *The State of Working Britain*, Manchester: Manchester University Press, pp. 29–46.

Janoski, H. (1994) *Comparative Political Economy of the Welfare State*, Cambridge: Cambridge University Press.

Jensen, L. and Chitose, Y. (1997) Will Workfare Work? Job Availability for Welfare Recipients in Rural and Urban America, *Population Research and Policy Review 16*, pp. 383–395.

Jessop, B. (1994) Post-Fordism and the State. In A. Amin (Ed.), *Post-Fordism: A Reader*, Oxford: Blackwell, pp. 251–279.

Jessop, B. (1997) Capitalism and its Future: Remarks on Regulation, Government and Governance, *Review of International Political Economy 4*, 3, pp. 561–581.

Jessop, B. (1999) The Changing Governance of Welfare: Recent Trends in its Primary Functions, Scale, and Modes of Coordination, *Social Policy and Administration, 33*, 4, pp. 348–359.

Jessop, B. (2002) Time and Space in the Globalization of Capital and their Implications for State Power, *Rethinking Marxism 14*, 1, pp. 97–117.

Jessop, B. (2003) From Thatcherism to New Labour: Neo-liberalism, Workfarism and Labour Market Regulation, Dept. of Sociology University of Lancaster, at http://www.comp.lancs.ac.uk/ sociology/soc131rj.pdf.

Johnson, C. and Tonkiss, F. (2002) The Third Influence: The Blair Government and Australian Labor, *Policy and Politic 30*, 1, pp. 5–18.

Johnson, P. (2001) New Labour: A Distinctive Vision of Welfare Policy? In S. White (Ed.), *New Labour – The Progressive Future?* Basingstoke: Palgrave, pp. 63–76.

Jones, M. (2000) *New Institutional Spaces: TECs and the Remaking of Economic Governance*, London: Jessica Kingsley.

Jones, M. and Gray, A. (2001) Social Capital, or Local Workfarism? Reflections on Employment Zones, *Local Economy 16*, 2, pp. 178–186.

Keep, E. (2000) *Creating a Knowledge-Driven Economy: Definitions, Challenges and Opportunities*, SKOPE Policy Paper, Warwick: University of Warwick.

Kellard, K. (2002) Job Retention and Advancement in the UK: A Developing Agenda, *Benefits 34, 10, 2*, pp. 93–98.

King, D. (1995) *Actively Seeking Work? The Politics of Unemployment and Welfare Policy in the United States and Great Britain*, Chicago: Chicago University Press.

King D. and Wyckham-Jones, M. (1999) Bridging the Atlantic: The Democratic (Party) Origins of Welfare to Work, in M. Powell (Ed.), *New Labour, New Welfare State*, Bristol: Polity Press, pp. 257–280.

Kleinman, M. and West, A. (1998) Employability and the New Deals, *Local Economy 5*, 3, pp. 174–179.

Kraft, K. (1998) An Evaluation of Active and Passive Labour Market Policy, *Applied Economics 30*, pp. 783–793.

Krugman, P. (1994) *The Age of Diminished Expectations*, Cambridge, Mass.: MIT Press

Krugman. P. (2002) For Richer: How the Permissive Capitalism of the Boom Destroyed American Equality, *New York Times Magazine*, 20 October.

Kuhn, T.S. (1962) *The Structure of Scientific Revolutions*, Chicago: University of Chicago Press.

Larsen, C. (2002) Policy Paradigms and Cross-National Policy (Mis)Learning from the Danish Employment Miracle, *Journal of European Public Policy 9*, 5, pp. 715–735.

Laville J.-L. (1996) *L'économie solidaire, une perspective internationale*, Paris: Desclée de Brouwer, collection 'sociologie économique'.

Lawrence, R. (1996) *Single World: Divided Nations? International Trade and the OECD Labour Markets*, Paris: OECD.

Layard, R. (1997a) *What Labour Can Do*, London: Warner Books.

Layard, R. (1997b) Preventing Long-Term Unemployment, in J. Philpott (Ed.), *Working for Full Employment*, London: Routledge, pp. 190–203.

Layard, R. (2001) *Welfare to Work and the New Deal*, London: Centre for Economic Performance.

Layard, R. (2004) *Good Jobs and Bad Jobs*, London: Centre for Economic Performance Occasional Paper 19.

Layard, R. and Nickell, S. (1987) The Labour Market, in R. Layard and R. Dornbusch (Eds.), *The Performance of the British Economy*, Oxford: Clarendon Press.

Layard, R., Nickell, S. and Jackman, R. (1991) *Unemployment: Macroeconomic Performance and the Labour Market*, Oxford: Oxford University Press.

Lewis, J. and Walker, R. (2000) What Works in the Delivery of Welfare to Work? in C. Chitty and G. Elam (Eds.), *Evaluating Welfare to Work, Papers from the Joint Department of Social Security and National Centre for Social Research Seminar held on June 16th 1999*, London: Social Security Research Report No. 67, pp. 7–32.

Liebfried, S. (2000) National Welfare States, European Integration and Globalization: A Perspective for the Next Century, *Social Policy and Administration 34*, 1, pp. 44–63.

Liebfried, S. (2001) *Welfare State Futures*, Cambridge: Cambridge University Press.

Lindsay, C. (2002) Long-Term Unemployment and the 'Employability Gap': Priorities for Renewing Britain's New Deal, *Journal of European Industrial Training* 26, 9, pp. 411–419.

Lindsay, C., McCracken, M. and McQuaid, R. (2003) Unemployment Duration and Employability in Remote Rural Labour Markets, *Journal of Rural Studies 19*, pp. 187–200.

Lindsay, C. and Mailand, M. (2004) Different Routes, Common Directions? Activation Policies for Young People in Denmark and the UK, *International Journal of Social Welfare 13*, pp. 195–207.

Lindsay, C. and Sturgeon, G. (2003) Local Responses to Long-Term Unemployment: Delivering Access to Employment in Edinburgh, *Local Economy 18*, 2, pp. 159–173.

Lisbon Group (1995) *Limits to Competition*, Cambridge, Mass: MIT Press.

Lister, R. (1999) To Rio via the Third Way: New Labour's Welfare Reform Agenda, *Renewal 8*, 4, pp. 9–20.

Lister, R. (2001) Doing Good by Stealth: The Politics of Poverty and Inequality under New Labour, *New Economy 8*, 2, pp. 65–70.

Lister, R. (2002) Citizenship and Changing Welfare States, in J.G. Andersen and P.H. Jensen (Eds.), *Changing Labour Markets, Welfare Policies and Citizenship*, Bristol: Policy Press, pp. 39–58.

Lloyd, C. and Payne, J. (2002a) On the 'Political Economy of Skill': Assessing the Possibilities for a Viable High Skills Project in the United Kingdom, *New Political Economy 7*, 3, pp. 367–395.

Lloyd, C. and Payne, J. (2002b) Developing a Political Economy of Skill, *Journal of Education and Work 15*, 4, pp. 365–390.

Lloyd, C. and Payne, J. (2003) What is the 'High Skills Society'? Some Reflections on Current Academic and Policy Debates in the UK, *Policy Studies 24*, 2/3, pp. 115–133.

Local Government Association (2001) *Beyond the New Deal: A New Approach to Local Employment Policy*, London: Local Government Association.

Lødemel, I. (2001) Discussion: Workfare in the Welfare State, in I. Lødemel and H. Trickey (Eds.), *'An Offer you Can't Refuse': Workfare in International Perspective*, Bristol: Policy Press, pp. 295–343.

Lødemel, I. and Trickey, H. (2001) *An Offer you Can't Refuse: Workfare in International Perspective*, Bristol: Policy Press.

Mackinnon, D. (2000) Managerialism, Governmentality and the State: A Neo-Foucauldian Approach to Local Economic Governance, *Political Geography 19*, pp. 293–314.

Marshall, B. and MacFarlane, R. (2000) *The Intermediate Labour Market: A Tool for Tackling Long-Term Unemployment*, York: Joseph Rowntree Foundation.

Martin, J. (2000) *What Works among Active Labour Market Policies: Evidence from OECD Countries' Experiences*, Paris: OECD Economic Studies 30.

Martin, R.L. (1988) The Political Economy of Britain's 'North–South Divide', *Transactions of the Institute of British Geographers NS, 13*, pp. 389–418.

Martin, R.L. (1993) 1993 Remapping Regional Policy: The End of the North–South Divide? *Regional Studies (Debates and Reviews) 27*, 8, pp. 797–805.

Martin, R.L. (1997) Regional Unemployment Disparities and their Dynamics, *Regional Studies 31*, pp. 35–50.

Martin, R.L. (1998) Regional Dimensions of Europe's Unemployment Crisis, Chapter 1 in P. Lawless, R.L. Martin and S. Hardy (Eds.), *Unemployment and Social Exclusion: Landscapes of Labour Market Inequality*, London: Jessica Kingsley, pp. 1–34.

Martin, R.L. (2001) Local Labour Markets: Their Nature, Performance and Regulation, Chapter 19 in G. Clark, M. Gertler and M. Feldman (Eds.), *Handbook of Economic Geography*, Oxford: Oxford University Press, pp. 455–476.

Martin, R. L. (2004) The Contemporary Debate over the North–South Divide: Images and Realities of Regional Inequality in Late Twentieth Century Britain, in A.R.H. Baker and M.D. Billinge (Eds.), *Geographies of England: The North South Divide: Imagined and Material*, Cambridge: Cambridge University Press, pp. 15–43.

Martin, R.L. and Morrison, P.S. (2003) Thinking about the Geographies of Labour, in R.L. Martin and P.S. Morrison (Eds.), *Geographies of Labour Market Inequality*, London: Routledge, pp. 3–20.

Martin, R.L., Nativel, C. and Sunley, P. (2001) The Local Impact of the New Deal: Does Geography Make a Difference, in R. Martin and P. Morrison (Eds.), *Geographies of Labour Market Inequality*, London: TSO.

Martin, R.L. and Sunley, P. (1999) Unemployment Flow Regimes and Regional Unemployment Disparities, *Environment and Planning A, 31*, pp. 523–550.

Mauss, M. (1990) *The Gift: The Form and Reason for Exchange in Archaic Societies*, London: Routledge.

McQuaid R. and Lindsay, C. (2002) The 'Employability Gap': Long Term Unemployment and Barriers to Work in Buoyant Labour Markets, *Environment and Planning C: Government and Policy 20*, pp. 613–628.

McVicar, D. and Podivinsky, J. (2003) *How Well Has the New Deal for Young People Worked in the UK Regions?* Northern Ireland Economic Research Paper Number 79.

Mead, L. (1997) *From Welfare to Work: Lessons from America*, A. Deacon (Ed.), London: Institute of Economic Affairs, Choice in Welfare Number 39.

Meager, N. and Evans, C. (1998) *The Evaluation of Active Labour Market Measures for the Long Term Unemployed*, Geneva: ILO Employment and Training Papers 16.

Meijers, F. and Riele, K. Te (2004) From Controlling to Constructive: Youth Unemployment Policy in Australia and the Netherlands, *Journal of Social Policy 33*, 1, pp. 3–25.

Millar, J. (2000a) *Keeping Track of Welfare Reform: The New Deal Programmes*, York: Joseph Rowntree Foundation.

Millar, J. (2000b) Lone Parents and the New Deal, *Policy Studies 21*, 4, pp. 333–345.

Millar, J. and Evans, M. (2003) *Lone Parents and Employment: An International Comparison of What Works*, London: DWP 181.

Mohan, J. (2000) New Labour, New Localism? *Renewal 8*, pp. 56–62.

Mulgan, R. (2000) Accountability: An Ever-Expanding Concept?, *Public Administration 78*, 3, pp. 555–573.

Murray, C. (1984) *Losing Ground*, New York: Basic Books.

Nathan, M. (2000) *In Search of Work: Employment Strategies for a Risky World*, London: Industrial Society.

Nathan, M., Simmonds, D. and Ward, M (1998) *Joining Up? The New Deal, the Public Sector and the Employer Option*, Manchester: Centre for Local Economic Strategies.

National Employment Panel (2004) *A New Deal for All, Report of the Working Group on New Deal 25 Plus*, Chaired by Lord Adebowale, London: National Employment Panel.

Nativel, C. (2004) *Economic Transition, Unemployment and Active Labour Market Policy: Lessons and Perspectives from the East German Bundesländer*, Birmingham: Birmingham University Press.

Nativel, C., Sunley, P. and Martin, R. (2003) Localising Welfare-to-Work? Territorial Flexibility and the New Deal for Young People, *Environment and Planning C: Government and Policy 20*, pp. 911–932.

Newman, K. and Lennon, C. (1995) *Finding Work in the Inner City: How Hard Is it Now? How Hard Will it Be for AFDC Recipients?* Working Paper 76, New York: Russell Sage.

Nickell, S. (2003) A Picture of European Unemployment: Success and Failure, *Discussion Paper 577*, London: Centre for Economic Performance.

Nickell, S. (2004) Poverty and Worklessness in Britain, *Economic Journal 114*, C1–C25.

Obstfeld, M. and Peri, G. (2004) Regional Nonadjustment and Fiscal Policy: Lessons for EMU, in G. Hess and E. Van Wincoop (Eds.), *Intranational Macroeconomics*, Cambridge: Cambridge University Press, pp. 221–271.

Ochel, W. (2004) *Welfare to Work Experiences with Specific Work-First programmes in Selected Countries*, CES IFO Working Paper no. 1153, Munich: IFO.

ODPM: Housing, Planning, Local Government and the Regions Committee (2003) The Effectiveness of Government Regeneration Initiatives, Seventh Report of Session 2002–3, Vol. 1, House of Commons 76–I.

OECD (1983) *Employment Outlook*, Paris: OECD.

OECD (1988) *Measures to Assist the Long-Term Unemployed: Recent Experience in Some OECD Countries*, Paris: OECD.

OECD (1994) *The Jobs Study, Volumes 1 and 2*, Paris: OECD.

OECD, (1998) *Local Management for More Effective Employment Policies*, Paris: OECD.

OECD (1999) *The Local Dimension of Welfare-to-Work: An International Survey*, Paris: OECD.

OECD (2000) Rewarding Work, *Employment Outlook*, June, pp. 7–10.

OECD (2001) *Local Partnerships for Better Governance*, Paris: OECD.

OECD (2002) *Employment Outlook*, Paris: OECD

OECD (2003) *Managing Decentralisation: A New Role for Labour Market Policy*, Paris: OECD.

Office of National Statistics (2003) Labour Force Survey 2003, ONS: London.

Oliker, S. (1994) Does Workfare Work? Evaluation Research and Workfare Policy, *Social Problems 41*, 2, pp. 195–213.

Okun, A. (1975) *Equality and Efficiency: The Big Trade-off*, Washington: Brookings Institute.

Payne, J. and Keep, E. (2003) Re-visiting the Nordic Approaches to Work Re-organization and Job Redesign: Lessons for UK Skills Policy, *Policy Studies 24*, 4, pp. 205–225.

Peck, J. (1994) Regulating Labour: The Social Regulation and Reproduction of Local Labour Markets, Chapter 7 in A. Amin and N. Thrift (Eds.), *Globalization, Institutions and Regional Development in Europe*, Oxford: Oxford University Press, pp. 147–176.

Peck, J. (1996) *Work-Place: The Social Regulation of the Labour Market*, New York: Guildford Press.

Peck, J. (1998a) Workfare in the Sun: Politics, Representation, and Method in U.S. Welfare-to-Work Strategies, *Political Geography 17*, pp. 535–566.

Peck, J. (1998b) Workfare: A Geopolitical Etymology, *Environment and Planning D; Society and Space 16*, pp. 133–161.

Peck, J. (1999) New Labourers? Making a New Deal for the 'Workless Class', *Environment and Planning C: Government and Policy 17*, pp. 345–372.

Peck, J. (2001) *Workfare States*, New York: Guildford Press.

Peck, J. and Theodore, N. (1998) The Business of Contingent Work: Growth and Restructuring in Chicago's Temporary Employment Industry, *Work, Employment, Society 12*, 4, pp. 655–674.

Peck, J. and Theodore, N. (2000a) 'Work First': Workfare and the Regulation of Contingent Labour Markets, *Cambridge Journal of Economics 24*, pp. 119–138.

Peck, J. and Theodore, N. (2000b) Beyond 'Employability', *Cambridge Journal of Economics 24*, pp. 729–749.

Peck J. and Theodore, N. (2001) Exporting Workfare/Importing Welfare-to-Work: Exploring the Politics of Third Way Policy Transfer, *Political Geography 20*, pp. 427–460.

Peck, J. and Tickell, A. (2002) Neoliberalizing Space, in N. Brenner and N. Theodore (Eds.), *Spaces of Neoliberalism: Urban Restructuring in North America and Western Europe*, Oxford: Blackwell, pp 33–57.

Philpott, J. (1999) After the Windfall: The New Deal at Work, *IEA Economic Affairs*, September, pp. 14–19.

Pierson, P. (2001) *The New Politics of the Welfare State*, Oxford: Oxford University Press.

Pigou, A.C. (1933) *Theory of Unemployment*, London: Macmillan.

Pinch, S. (1997) *Worlds of Welfare*, London: Routledge.

Plant, R. (2003) Citizenship and Social Security, *Fiscal Studies 24*, 2, pp. 153–166.

Powell, M. (2002) (Ed.) *Evaluating New Labour's Welfare Reforms*, Bristol: Policy Press.

Prideaux, S. (2001) New Labour, Old Functionalism: The Underlying Contradictions of Welfare Reform in the US and UK, *Social Policy and Administration 35*, 1, pp. 85–115.

Ramia, G. and Carney, T. (2001) Contractualism, Managerialim and Welfare: The Australian Experiment with a Marketised Employment Services Network, *Policy and Politics 29*, 1, pp. 59–80.

Reich, R. (2001) *The Future of Success*, New York: Alexander Knopf.

Reimer, S. (2003) Employer Strategies and the Fragmentation of Local Employment: The Case of Contracting out Local Authority Services, in R.L. Martin and

P.S. Morrison (Eds.), *Geographies of Labour Market Inequality*, London: Routledge, pp. 110–128.

Riley, R. and Young, G. (2000) *The New Deal for Young People: Implications for Employment and the Public Finances*, London: NIESR, ESR62.

Ritchie, J. (2000) New Deal for Young People: Participants Perspectives, *Policy Studies 21*, 4, pp. 301–312.

Robinson, D. (1986) *Monetarism and the Labour Market*, Oxford: Clarendon Press.

Robinson, P. (2000) Active Labour-Market Policies: A Case of Evidence Based Policy-Making? *Oxford Review of Economic Policy 16*, 1, pp. 13–26.

Rodger, J., Burniston, S. and Lawless, M. (2000) *New Deal for Young People: Delivery and Performance in Private Sector Lead Areas*, Sheffield: Employment Service Research and Development Report ESR53.

Rowthorn, R. (2000) Kalecki Centenary Lecture: The Political Economy of Full Employment in Modern Britain, *Oxford Bulletin of Economics and Statistics 62*, 2, pp. 413–425.

Rubery, J. and Wilkinson, F. (1994) *Employer Strategy and the Labour Market*, Oxford: Oxford University Press.

Rubery, J., Earnshaw, J., Marchington, M., Cooke, F.L. andVincent, S. (2002) Changing Organizational Forms and the Employment Relationship, *Journal of Management Studies 39*, 5, pp. 645–572.

Sargeant, G. and Whiteley, P. (2000) *A Good Deal Better*, London: Industrial Society.

Scharpf, F. (2001) The Viability of Advanced Welfare States in the International Economy: Vulnerabilities and Options, *European Review 8*, 3, pp. 399–425.

Schmidtz, D. and Goodin, R.E. (1998) *Social Welfare and Individual Responsibility: For and Against*, Cambridge: Cambridge University Press.

Schoenberger, E. (2000) *The Cultural Crisis of the Firm*, Oxford: Blackwell.

Sennett, R. (1998) *The Corrosion of Character: The Personal Consequences of Work in the New Capitalism*, New York: Norton.

Seuret, F. (2001) Emplois-jeunes: Cinq ans, et après?, *Alternatives Economiques 190*, March 2001, pp. 46–51.

Smith, G. (1999) *Area-based Initiatives: The Rationale and Options for Area Targeting*, CASE paper 25, London: CASE.

Smith A., (2004) Welfare to Work Package Builds on Success, DWP Press Release, 17 March.

Snape, D. (1998) *New Deal for Young Unemployed People: A Good Deal for Employers? Findings from Preliminary Qualitative Research*, London: DWP 6.

Social Exclusion Unit (1998) *Bringing Britain Together: A National Strategy for Neighbourhood Renewal*, London: Cabinet Office, Cm 4045.

Social Exclusion Unit (2004) *Jobs and Enterprise in Deprived Areas*, London: Office of the Deputy Prime Minister.

Solow, R.M. (1998) *Work and Welfare*, Princeton: Princeton University Press.

Stewart, M. (1999) Low Pay in Britain, in P. Gregg and J. Wadsworth (Eds.), *The State of Working Britain*, Manchester: Manchester University Press, pp. 225–248.

Storper, M. (1996) *The Regional World: Territorial Development in a Global Economy*, New York: Guildford Press.

Sumaza, C. (2001) Lone Parent Families within New Labour Welfare Reform, *Contemporary Politics* 7, 3, pp. 231–247.

Sunley, P., Martin, R. and Nativel, C. (2001) Mapping the New Deal: The Local Disparities in the Performance of Welfare-to-Work, *Transactions of the Institute of British Geographers 26*, pp. 484–512.

Surel, Y. (2000) The Role of Cognitive and Normative Frames in Policy Making, *Journal of European Public Policy* 7, 4, pp. 495–512.

Swank, D. (2002) *Global Capital, Political Institutions, and Policy Change in Developed Welfare States,* Cambridge: Cambridge University Press.

Tamkin, P., Hillage, J. and Gerova, V. (2005) The Regional Implementation of the Employer Training Pilot in the UK, in OECD (Ed.), *Rising Expectations: New Perspectives for the Low-Skilled,* Paris: OECD (forthcoming).

Tavistock Institute (1999) *A Review of Thirty New Deal Partnerships,* Sheffield: Employment Service Research and Development Report ESR32.

Taylor Gooby, P. (2001) *Welfare States under Pressure,* London: Sage.

Theodore, N. and Peck, J. (1999) Welfare-to-Work: National Problems, Local Solutions? *Critical Social Policy 19,* 4, pp. 485–510.

Torfing, J. (1999) Workfare with Welfare: Recent Reforms of the Danish Welfare State, *Journal of European Social Policy 9,* 1, pp. 5–28.

Training Standards Council (2000) *New Deal 18–24 Inspection Report November 1999 Birmingham Unit of Delivery.*

Trickey, H. and Walker, R. (2001) Steps to Compulsion within British Labour Market Policies, in I. Lødemel and H. Trickey (2001) (Eds.), *'An Offer you Can't Refuse': Workfare in International Perspective,* Bristol: Policy Press, pp. 181–214.

Turner, D. (2004) Schemes Shift Jobseekers to Other Benefits, *Financial Times,* 26 April 2004, p. 3.

Turok, I. and Edge, N. (1999) *The Jobs Gap in Britain's Cities,* Bristol: Policy Press and Joseph Rowntree Foundation.

Turok, I. and Webster, D. (1998) The New Deal: Jeopardised by the Geography of Unemployment, *Local Economy,* pp. 1–20.

Van Oorschot, W. and Abrahamson, P. (2003) The Dutch and Danish Miracles Revisited: A Critical Discussion of Activation Policies in Two Small Welfare States, *Social Policy and Administration 37,* 3, pp. 288–304.

Van Reenen, J. (2001) *No More Skivvy Schemes? Active Labour Market Policies and the British New Deal for the Young Unemployed in Context,* London: Institute for Fiscal Studies.

Walker, R. (1991) *Thinking about Workfare: Evidence from the USA,* London: HMSO.

Walker, R. and Wiseman, M. (2003) Making Welfare Work: UK Activation Policies under New Labour, *International Social Security Review 56,* 1, pp. 3–29.

Walsh, K. Atkinson, J. and Barry, J. (1999) *The New Deal Gateway: A Labour Market Assessment,* ES Research and Development Report, ESR24.

Webster, D. (1996) The Simple Relationship between Long-Term and Total Unemployment and its Implications for Policies on Unemployment and Area Regeneration, *Glasgow City Housing Working Paper,* March.

Webster, D. (1997) The L-U Curve: On the Non-Existence of a Long-Term Claimant Unemployment Trap and its Implications for Employment and Area Regeneration, *University of Glasgow Dept of Urban Studies Occasional Paper 36*, May.

Webster, D. (1999) Targeted Local Jobs: The Missing Element in Labour's Social Inclusion Policy, *New Economy 6*, pp. 193–198.

Webster, D. (2000) The Geographical Concentration of Labour-Market Disadvantage, *Oxford Review of Economic Policy 16*, 1, pp. 114–127.

Webster, D. (2003) Long-Term Unemployment, the Invention of Hysteresis and the Misdiagnosis of the UK's Problem of Structural Unemployment, Paper given at Cambridge Journal of Economics Conference 17–19 September 2003, Economics for the Future.

White, M. (2000) New Deal for Young People: Towards an Ethical Employment Policy? *Policy Studies 21*, 4, pp. 285–299.

White, M. and Riley, R. (2002) *Findings from the Macro Evaluation of the New Deal for Young People*, London: DWP 168.

Wilkinson, D. (2003) *New Deal for People Aged 25 and Over: A Synthesis Report*, London: DWP 161.

Willettts, D., Hillman N. and Bogdanor A. (2003) *Left out, Left Behind: The People Lost to Britain's Workforce*, London: Policy Exchange.

Wood, A. (1994) *North–South Trade, Employment and Inequality: Changing Fortunes in a Skill-Driven World*, Oxford: Clarendon.

Woodfield, K., Bruce, S. and Ritchie, J. (2000) *New Deal for Young People: the National Options. Findings from a Qualitative Study among Individuals*, Sheffield: Employment Service Research and Development Report ESR37.

Work and Pensions Committee (2002) The Government's Employment Strategy, Minutes of Evidence Wednesday 1 May 2002, House of Commons 815–I, Session 2001–2, London: TSO.

Wright, S. (2002) Activating the Unemployed: The Street-Level Implementation of UK Policy, in J. Clasen (Ed.), *What Future for Social Security?: Debates and Reforms in National and Cross-National Perspective*, Bristol: Policy Press, pp. 235–250.

Wright, S., Kopac, A. and Slater, G. (2004) Continuities within Paradigmatic Change: Activation, Social Policies and Citizenship in the Context of Welfare Reform in Slovenia and the UK, *European Societies 6*, 4, pp. 511–534.

# Index

Printed and bound by CPI Group (UK) Ltd, Croydon, CR0 4YY

27/10/2024

14580382-0001